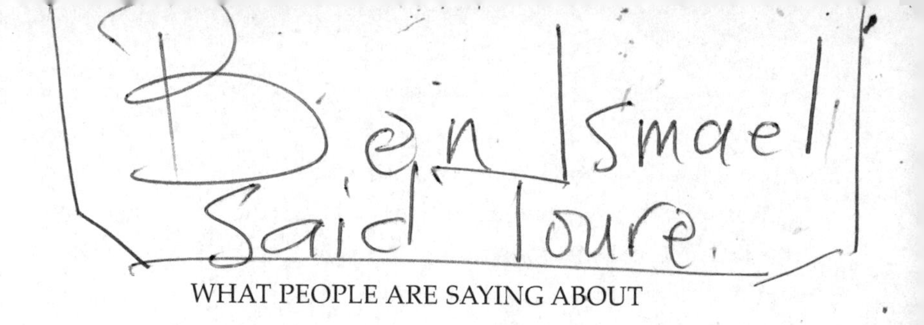


WHAT PEOPLE ARE SAYING ABOUT

# LIVING JAINISM



A remarkably comprehensive ( ... ly of the interconnectedness of b ... ce for our approach to the environment dr ... the principles of interdependence and reverence for life. Truth is presented as not only one- but also many-sided, opening up interfaith dialog and an attitude of non-violence relevant to the modern human predicament. The book encourages us to develop relationships based on co-operation, compassion and trust. Jainism is presented as a philosophy to rein in the materialistic and exploitative trends of modern life and overcome our sense of separation from Nature and each other. The book is also hopeful of human possibilities and presents an expanded view of perception and logic. Essential reading for anyone interested in the philosophy and implications of Jainism for our time.
**David Lorimer**, Scientific and Medical Network

Another profound and beautiful book from Aidan Rankin and his co-author. Their exquisite scholarship and love of Jainism emanates from every page; and brings the wisdom and modern-day application of this ancient but highly relevant ecological spirituality to a modern readership. Their focus on 'personal ecology' is timely, compelling and accessible. I cannot recommend this book more highly.
**Rev Lynne Sedgmore**, CBE, CEO, 157 Group of FE Colleges

An outstanding book from two very talented scholars of Jain philosophy and wisdom. It tackles a very complex and important topic, demonstrating the vast depth of Jain science and metaphysics.
**Dr Atul K. Shah**, CEO, Diverse Ethics Ltd

# Living Jainism

## An Ethical Science

# Living Jainism

## An Ethical Science

Kanti V. Mardia

and

Aidan D. Rankin

Winchester, UK
Washington, USA

First published by Mantra Books, 2013
Mantra Books is an imprint of John Hunt Publishing Ltd., Laurel House, Station Approach,
Alresford, Hants, SO24 9JH, UK
office1@jhpbooks.net
www.johnhuntpublishing.com
www.mantra-books.net

For distributor details and how to order please visit the 'Ordering' section on our website.



ISBN: 978 1 78099 912 8



A CIP catalogue record for this book is available from the British Library.

Design: Stuart Davies

Printed and bound by CPI Group (UK) Ltd, Croydon, CR0 4YY

# CONTENTS

For my daughter Bela and my grandsons Ashwin and Sashin – K.V.M.

For Anne and David Rankin, for Brian Scoltock – A.D.R.

*Parasparopagraho Jivanam:* All life is bound together

# Foreword

Jainism is one of the most important spiritual and religious traditions in the world – and one of the oldest, with its origins going probably well back over 2500 years. Perhaps most notably, Jainism has a many-sided literature as extensive as that of any other of the world's great religions, including art, science and mathematics, though most of it has not been given its proper study or rightful place in world thought.

Though the number of its current adherents may be relatively small, somewhat like Judaism, Jainism has an influence far beyond its numbers. It has great philosophical and mystical traditions and its followers are usually well educated and highly cultured. Though it mainly has adherents in India, the orientation of Jainism has always been towards the whole of life, and it is now gaining respect throughout the world.

Yet Jainism has its own unique nature and teachings, which can be quite different from religion in the western sense of the term. Jainism is not a belief system, but rather a way of right relationship with the living universe that we are inherently part of. It is not based upon a theology of a Creator that we must all bow down to, but upon recognition of the sacred nature of all life. Jainism is a spiritual science or way of higher knowledge and should be treated as such, if we want to really understand and appreciate the great gifts of insight that it can bestow upon us.

Jainism is a religion of life and honoring life. For Jainism, life is perhaps more important and real than any concept of divinity. How we relate to life is our true religion, which implies how we act in our daily lives, not simply a seeking of salvation or going to a heavenly world after death. Jainism begins with the unity of life and asks us to discover the implications of that unity, and reorient our behavior accordingly.

Jainism takes this honoring of life into its interaction with other religions, and with human culture as a whole. Jainism does not spread by mere assumption of a belief, an emotion, or a change of outer identity. It helps us connect with the deeper powers of nature, which to a great extent must remain nameless and transcend any mere human language connotations. Jains do not proselytize or preach, but they do teach and do this by example and behavior rather than by promoting divisive or separatist doctrines.

Jainism places a great deal of emphasis on the conduct of right living, which means first of all respect for all living beings, more so than it does on promoting any system of thought or particular faith. For Jainism, ahimsa or non-violence precedes and is greater than any deity or act of worship, or even any outer seeking of knowledge. But true ahimsa does not simply mean avoiding overt acts of physical violence. It requires working to reduce the amount of harm, violence, division and conflict not only in the world but also in ourselves. Ahimsa means non-interference with the natural balance and allowing every aspect of life and nature its own place and dignity.

This means that Jainism is eminently an ecological religion, woven into the web of life. It teaches us how to honor the ecology around us, not simply as a naturalistic phenomenon but as part of a sacred reality. This is not just a matter of respecting the way in which nature works at an outer level. More than that, it implies a deep reverence for all creatures as the most basic form of worship. It rests upon allowing each aspect of nature its sacred space; a place where we do not interfere with it, but allow it to flower naturally and to its fullest capacity.

Jainism also teaches us how to manage our personal ecology, which is how we conduct our personal lives, especially the psychological environment that we create and sustain by our thoughts and actions. We may refrain from actual violence but may still not be helping the world, if we are projecting thoughts of anger, fear or desire. Jainism implies right cultivation of our own individual nature, of both body and mind, and this includes respect for how they work and not interfering with their organic function. This extends to our attitude to food, which is our most basic interaction with life, and this should first of all be free of violence.

For the Jain mind the soul is everywhere and there are innumerable souls throughout all of nature, including at very minute levels. Intelligence is part of the fabric of nature, not a human invention or social production. Our actions accumulate karmic matter that can cause harm and suffering, but can also be cleansed away by right thought and right action. There is a logic and a science as to how we can understand the karmas we are creating and reduce all negative karmas, like toxins, from our inner being. In this way, free of limitations and impurities, we can know the highest truth of life and consciousness.

The authors of *Living Jainism*, Professor Kanti Mardia, DSc, and Aidan Rankin, PhD, have made Jainism alive, practical and relevant for the modern reader. They have made clear the traditional tenets and practices of this ancient way of life, culture and spirituality. The book has a great deal of philosophical depth and detail; perhaps more so than any available book on Jainism in the English language. It presents one of the most cogent, analytical approaches to understanding the law of karma, not simply as a set of moral imperatives, but as a code of universal laws and ethics, and as actual energies that we need to harmonize, which is one of the core insights of Jainism.

*Living Jainism* introduces 'Jain Science' or the 'Jain way of knowledge' with depth and precision, opening up the vistas of the enlightened Jain mind to help inspire us to pursue the higher knowledge of the universe. *Living Jainism* serves as a good modern textbook on Jainism, though it is a book that also has practical relevance and can be appreciated by interested readers from diverse backgrounds. It should serve to open the mind of any deep or sincere reader to the treasures that Jainism can still offer the world.

Certainly Jainism has many gifts and much wisdom that is essential and crucial for dealing with our current era of major ecological crisis that is only likely to grow in years to come. How we may define this crisis is not that important. But we must recognize that we as human beings have upset the natural balance on the planet, and need to work hard to restore this both for the well-being of the planet and that of ourselves. For this lofty purpose, we need a change of values and life-style, not mere argument over theology, politics or economics. We need not abandon science but we must let go of, or at least reduce, selfishness and greed. Jain thought and the Jain way of life can provide helpful guidance in this process.

Jainism can help us get beyond our current global crisis and in a way that is respectful to all life. For those serious about studying Jainism today, *Living Jainism* is a good place to start. Living Jainism, which means living with truth, simplicity and respect for all, will certainly protect, promote and foster all of life and help bring in a new era of peace for our turbulent world.

David Frawley, PhD, author, *Yoga and Ayurveda, Tantric Yoga, Wisdom of the Ancient Seers,* Director, American Institute of Vedic Studies (www.vedanet.com)

# Acknowledgements

This book has been made possible by the generosity of the Mardia Punya Trust, whose Trustees provided moral support and calm, critical guidance at every stage of the writing. It could not have been written without the scholarly inspiration of Professor Padmanabh S. Jaini of the University of California, Berkeley. Professor Jaini has played a seminal role in opening western and eastern eyes alike to Jain history and culture, bringing it into the mainstream of academic study and public appreciation.

We would like to thank the publisher, John Hunt, for his faith in and enthusiasm for the book and the ideas it contains. Raj Khullar, Pavan Mardia and Neeta Mehta read the manuscript and made helpful comments. Alice Hamar gave us invaluable help with the manuscript as well as detailed comments for which we shall always be grateful. We are also grateful to Derek Marshall and Mike Hall for their help with the illustrations in the book.

Finally, we thank members of the Yorkshire Jain Foundation for their kindness, warm hospitality and continuous encouragement.

K.V.M.
A.D.R.

## Special Note

Many sections of this book are a development and progression of ideas and materials first explored in Mardia, K.V., *The Scientific Foundations of Jainism* (1990, 2007, Motilal Banarsidass, Delhi). In this earlier book, the idea of Jainism's 'Four Axioms' was first propounded. Our present book builds on this theme, presenting the Axioms as 'Four Noble Truths' and making the concepts readily accessible to the general reader. It also contains additional material not covered by the earlier book, especially in relation to Jain culture and society and the relevance of Jain thought to scientific and social questions. The two books can be read independently of each other.

# Introduction

*I adore so greatly the principles of the Jain religion that I would like to be reborn in a Jain community.*
Attributed to George Bernard Shaw, playwright and thinker, 1856-1950 CE[1]

*Non-violence and kindness to living beings is kindness to oneself.*
Mahavira, Twenty-Fourth *Tirthankara* (Path-Finder) of Jainism, 559-527 BCE[2]

The two quotations with which we begin this study bind together two worlds that have for too long been considered discrete, incompatible, and at best parallel spheres. That this separation is to the detriment of both worlds is now widely accepted across the intellectual spectrum and even in the political arena. Our purpose here is not to add to the larger chorus of lamentations about the sense of loss experienced by the spiritual and scientific communities alike. Nor (except in passing and where relevant) do we explore the origins of that loss, or how the idea of a dichotomy between the 'scientific' and 'spiritual' worlds arose. Much has already been written and discussed on this theme, but for the most part it lies outside the scope of our study.

## Our Approach

Instead, our aim is in some respects more practical: the healing of a wounding breach between two of the most powerful areas of knowledge, one based on reason and logic, the other on intuition and wonder. At a deeper level, we wish to show that this division is illusory. There is in reality a natural kinship between the spiritual inquirer and the scientist. Both are questing for underlying truths about the universe and are inspired by a similar sense of curiosity and awe. The scientific and the spiritual researcher alike are stretching the boundaries of, and sometimes transcending altogether, those systems of thought that have ceased to represent accumulated wisdom, which is eternally supple, and instead solidified into dogmas. 'Science', for our purpose, encompasses what we have learned to call the physical sciences. At the same time, we define the word in its broader and more original form, derived from the Latin *scire,* to know. The scientific quest is the pursuit of knowledge in its entirety, which includes an understanding of philosophy and psychology, artistic and creative impulses or those aspects of existence and experience that cannot be measured or quantified. 'Spiritual' we define, equally broadly, as the domain beyond immediate or conventional understanding, principally the hidden connections that exist within the universe as a whole and the planet we inhabit.

The approach behind this book is, to use a fashionable word, *holistic.* It is about the blurring or abolition of boundaries, about making connections in place of the convenient academic habit of compartmentalizing. In short, we are interested in continuities rather than divisions. From this arises the concept of *spiritual science,* or a continuum of reason and intuition, metaphysical speculation and the accumulation of material or 'solid' facts. Jain philosophy is part of the complex of ideas known as 'Indic' thought, because they originate in India and share common characteristics as well as distinct identities. Indic thought therefore encompasses Hindu and Buddhist teachings as well as Sikhism and Jain Dharma. The scientific-spiritual continuum has been one of the constants of Indic thought.

The special resonance of Jain concepts today arises in part from their familiarity rather than their novelty. In the centrality it ascribes to the autonomous individual as a unique being endowed with inalienable rights, inherent value, unlimited potential and the ability to make choices, Jain teachings accord well with the patterns that have resonated throughout western thought. These include the Judaic and Christian traditions, secular democratic thought with its roots in classical antiquity and the finest Enlightenment of continuous questioning and improvement. Jainism differs from almost all prevailing Buddhist and Vedic Hindu currents of opinion in that individuality is not cast aside but perfected by the process of spiritual enlightenment. It also explicitly regards the pursuit of all knowledge, all forms of truth, as part of the spiritual quest.

From this point of almost reassuring likeness to the West's dominant values, Jain thought then challenges those values at a more radical level, in the literal sense of *from the roots upwards*, than arguably any other spiritual force. Its concept of the individual is larger, more 'inclusive' than that which we have come to accept. First, it encompasses everything that contains life. This is more than a sensibility (although it includes this): it is awareness that all of life is part of a continuum and that all beings, however 'primitive' or 'undeveloped' they might look to human eyes, have a unique purpose and the potential for enlightenment. The Earth and the universe were, according to the early Jain sages, teeming with life. Even life forms invisible to the naked eye can be highly complex systems that play an unseen but critical role in sustaining life in its entirety. No life form is insignificant, indeed all lives are critically important.

Western science is bearing out more and more the conclusions of Jain thinkers, arrived at through investigation and intuition: classic scientific method. The Jain concept of an interdependent 'hierarchy of life' accords well with the secular scientific understanding of evolution, giving it a spiritual underpinning. Yet Jain awareness of life in its full variety and complexity invites us to ask radical questions about our expectations as human beings and our relationship with 'the rest' of nature. Jain principles, in short, make nonsense of the idea that our intelligence or creative powers permit us to exploit or exterminate supposedly 'lesser' species or treat the Earth's resources as unlimited and at our disposal. On the contrary, Jain principles enjoin human beings to live within limits precisely because our intelligence can allow us to understand that all living things are of value and worth, being reflections or aspects of ourselves. The concept of Careful Action lies at the core of Jain practice. It involves the use of our thought processes consciously to avoid or minimize all actions that harm or adversely impinge upon other forms of life. It is no exaggeration to say that the notion of an 'ecological footprint' has been part of Jain thinking for thousands of years, indeed from the first emergence of Jain culture. Out of this sense of the sacred (and highly practical) value of all life grows the idea that the individual seeking enlightenment (that is, true individuality) can only come close to this by co-operation and fellowship. Needless to say, such social solidarity goes well beyond the human realm and involves learning to work with rather than against the grain of nature. True human intelligence, true scientific understanding, is expressed by restraint rather than conquest, which is a sign of limited spiritual awareness and a sense only of short-term gain rather than long-term advantage.

The epigraphs that begin this chapter together provide an example of spiritual science and the potential that Jain philosophy has to influence the wider global culture's ways of thinking and living. This does not mean 'becoming' or even mimicking Jains, but listening to an ancient yet radical wisdom tradition and using it to think, criticize, evaluate and then act with care. Mahavira, about whom more will be said in the next chapter and Appendix 1, is the seer who gave Jainism the characteristics most widely recognized in today's world. His injunction to us, to understand the connections between living systems, is at the same time an ethical teaching about restrained behavior and compassion, an ecological treatise summarized in a few words and an expression of rational self-interest.

## Jain Tradition

Jainism is one of the oldest practiced spiritual traditions in the modern world. It traces its origins in the most ancient expressions of Indic thought. Mahavira, who is the main point of reference for Jains today, was a contemporary of Gautama Buddha, but he built upon foundations that had been laid over many centuries and which can be traced to a prehistoric or mythical Lawgiver. This antiquity is reflected in some of the ideas associated with Jainism that are now considered 'advanced'. Chief amongst these is the sense that all forms of life are interconnected, no matter how minuscule they might appear to the human gaze or how peripheral they might seem to the human experience. Moreover, the idea of interconnectedness encompasses all areas of life. Jains do not compartmentalize living systems into 'animal, vegetable or mineral' categories because they are more aware of the continuities between them than the differences. From this it follows that the Earth, the 'global environment' which sustains life in its rich variety, must be conserved and cherished rather than 'conquered' or defiled. Spiritual wisdom and ecological ethics are as one. They are reflections of Dharma, the eternal law of the universe. Every aspect of Jain life and thought arises from the principle of interconnectedness and seeks to apply its lessons.

The discovery of modern preoccupations – with the environment and cultural diversity, for example – in the earliest spiritual teachings comes as no surprise from a Jain standpoint. For all thought, all actions, all aspects of the universe, are held to be cyclical rather than linear. They 'come round again' constantly until full enlightenment is achieved. For a Jain, the interconnectedness of all life includes all the stages of evolution, which contribute to a sum total of knowledge and consciousness. Parallel to physical or biological evolution is a process of spiritual evolution by which the soul (pictured by Jains as a 'life monad' or unit of pure life) passes through myriad incarnations and experiences on the way to full knowledge of itself. In other words, it makes a cyclical journey until the pure form is achieved. This journey transcends the divisions of species, gender and evolutionary stage. Therefore, all forms of experience are considered, not necessarily as 'equal' or 'the same', but as valid aspects of a larger truth.

From this primal awareness spring, in turn, two of the most powerful currents in Jain thought. The first of these is Ahimsa, non-violence or more correctly *the avoidance of harm* to all forms of life. Lay men and women aim to put this principle into practice through carefully considering all their actions and thoughts and their possible consequences. Ascetic men and women enact the principle dramatically through avoiding injury to all forms of life, including insects and even creatures invisible to the naked eye. They renounce material possessions and live as simply and as closely to nature as possible because the pursuit of material possessions, whatever their outward benefits, invariably harms the psyche and leads to the exploitation of others. This way of life is a reflection of their Greater Vows (see Chapter 1) and it is also a reminder to their lay brothers and sisters that material satisfaction is only transient and that there is an inward way that points to freedom from constraints and short-term thinking. Thus although they live close to nature, the ascetics are in fact seeking to move beyond natural boundaries and conventional human impulses. It is well understood, however, by lay Jains, who see it as the next phase of their journey to self-knowledge. For a community noted in India and elsewhere for commercial success, the spiritual ideal is to possess next to nothing!

The second powerful current of Jain thought is the idea of 'Many-Sidedness' or *Anekantavada*. This philosophy posits that all opinions and perspectives represent aspects of the truth because they reflect different stages of the spiritual journey mentioned above. And within the full range of opinions are incorporated those of other species – plant and mineral 'count' as much as

animal or human. Each unit of life is individual and distinctive. It is a unique entity with its own contribution to make, and at the same time it is part of an 'inhabited universe' filled with all varieties of life.

This perspective makes each individual simultaneously aware of his or her importance and un-importance, power and limitations, knowledge and ignorance. It has an impact on the way we view and behave towards other species and how we treat the environment. More than that, it affects the way we view the range of human thought and engage with opinions that we regard as fundamentally flawed or 'wrong'. From the Jain perspective, even these opinions represent an aspect of the truth, however perverted that might have become. They need to be accepted for what they are, even if they are necessarily opposed or even fought.

In general, disagreements and divergences of opinion take a less extreme form. As a method of inquiry, Many-Sidedness involves respecting diversity, bridging divisions, negotiating, establishing common ground and shared values. As a solution or method of conflict resolution, its goal is *synthesis*. Multifarious strands of opinion, differing and opposed viewpoints are fused together so that they develop into something new and distinctive. Arguably, the Jains practiced interfaith dialog thousands of years before the term was coined. They could also be described as pioneers of 'multiculturalism'. It corresponds to the original idea of respecting and being willing to open oneself to a range of cultural influences and viewing them as containing potential spiritual riches. This process does not require renouncing our most formative influences or 'native cultures', but at once enlarging upon and gaining a greater understanding of them. The Jains, who form minority communities in both India and the Diaspora, have incorporated a wide range of cultural influences and practices whilst retaining a distinct and clear identity, which they neither compromise nor aggressively promote. Their principle and practice of Many-Sidedness has enabled their minority culture to survive, i.e. by simultaneously integrating and maintaining its identity.

The Jain tradition is made special by its powerful scientific bias and its emphasis on the pursuit of knowledge. Because omniscience is a property of the pure soul or life monad, the pursuit of knowledge in all its forms is part of the journey to liberation. Ignorance, by contrast, is associated with material attachment and negative karmic influences. The dichotomy between science and spiritual or religious practice that has been a feature of many human cultures has never occurred in Jainism. Indeed the resemblances between the work of the scientific researcher and the spiritual practitioner tend to be emphasized. Both are seeking objective truths but, being human and subjective, are arriving at aspects of the truth.

It follows that Jains have consistently placed a strong emphasis on education (notably of both men and women) and valued scholarship, whether scientific, religious or legalistic. The training of the mind is as important to Jain practitioners as meditation. It extends beyond amassing facts to developing a critical faculty and learning to question one's own ideas and (crucially) one's own motives. Jainism combines scientific reasoning and the belief in the possibility of progress with a sense of humility at the vastness of the universe and our relatively small place within it. This means that Jains believe that they have a responsibility, as rational men and women, to learn as much about the workings of the cosmos as is practically possible. By so doing, they learn more about themselves. At the same time, they are aware of how little human beings can actually *know*, however 'advanced' our species might become.

The belief in progress is thereby tempered by an awareness of our limitations and our need to respect alternative viewpoints. Proof of spiritual and scientific progress, for Jains, lies in renouncing power over other species, fellow humans and natural resources. Such powers are viewed as symptoms of ignorance and arrogance. They are forms of material attachment that entrap the human spirit and are, in any case, founded on illusion. The same is held to be true of religious or political (and sometimes scientific) dogmas that limit the capacity for self-criticism and lead to acts of harm. Clinging to power and clinging to dogma inhibit all forms of knowledge, especially self-knowledge. They are as un-scientific as they are un-spiritual.

## Scientific Perspective

In an age when science is increasingly seeing beyond the limits of linear, mechanistic thinking, the Jain thought process begins to resonate well beyond an Indic minority population. The scientific foundations of Jainism contain the rational, methodical approach that western science evolved towards from the Renaissance and Enlightenment through the industrial age. It was a way of looking at the world that revived many of the approaches of classical Greece and Rome. These ran parallel to critical developments in Jain philosophy – along with the evolution of the Vedic Dharma of the Hindus and the emergence of Buddhism.

At the same time, the scientific approach of the Jains has not become fixated on matter at the expense of spirit. An understanding of the philosophy of Ahimsa and Anekantavada is believed to assist in that evolutionary process. It might also, in the true spirit of Many-Sidedness, lay the foundations for a cross-cultural synthesis of western and Indic approaches.

The face masks and brushes used by monks to protect infinitesimal life forms can easily convey an impression of puritanical austerity. What they should convey instead is a life-affirming vision of non-violence coupled with a scientific understanding of the hidden connections within nature.

Scientists are ever-more aware of the critical role of (for most) inconceivably tiny particles such as the neutrinos first identified by the Austrian physicist Wolfgang Pauli in 1930. The interaction of such elementary particles with each other plays a crucial part in the stability and continuity of life, in ways that we are only beginning to understand through scientific method but which are recognized intuitively within Jain culture.

Neutrinos, which are immensely difficult to detect because they have no electrical charge and travel close to the speed of light, surely bear some resemblance to the karmic particles (or karmons) that are held by Jains to affect all areas of life and consciousness. The significance of these karmic particles will be discussed at length throughout our book, for they form the basis for so much Jain thought, including scientific awareness. Jain ascetics who protect elementary and invisible life forms are living out in a literal way Mahavira's dictum that 'Non-violence and kindness to living beings is kindness to oneself.' They are making the connection between the survival of each life form and the survival of *all* life, indeed the life principle itself. Secular science is rapidly converging with these spiritual intuitions. The decisive role of tiny organisms such as plankton in the Earth's climatic stability is recognized by oceanographers and climatologists and with it the campaign for ocean conservation gathers momentum. For a Jain ascetic, this would be no surprise, but an expression of the obvious!

An important reason why Jains are under-represented is that one of their culture's greatest strengths can be a weakness in a vocal, media-driven age. Jain culture is characterized by dignity and modesty. Its principles are lived out each day without assertiveness or propagandist preaching. A practicing Jain is all too aware of his or her possible shortcomings (as a non-omniscient being) and of the often rigorous standards of living in accordance with Dharma (natural law). Jain culture discourages judgment of others and encourages self-criticism. It is a contemplative rather than doctrinaire spiritual practice in which those who lead – the Acharya or spiritual teacher, for instance – do so by example. Put simply, deeds mean more than words. The Jain is also more interested in learning from or listening to other perspectives than proclaiming his own. Jain teachings are viewed as a resource for those who seek to live by them and those who come from other traditions who seek them out.

This quiet confidence is the opposite of the missionary approach of many other faiths and secular ideologies. It is a source of inner strength for Jain communities around the world, but it is not best suited (and nor should it be) to a public discourse based largely on assertion and controversy. As a result, Jain quietness is frequently confused with quiet-*ism,* which it emphatically is not. Few in the scientific community know what Jains actually believe and how they think, still less what relationship this could have with the problems affecting modern science.

For those who seek scientific and spiritual convergence, Jainism is somewhat of a jewel in the shade, and the purpose of this book is to bring it into the light. Our book is itself structured according to the method of Anekantavada. It synthesizes the work of two authors, two minds approaching Jain philosophy from different starting points, but which are now converging on the pathway. Kanti Mardia comes from a long Jain lineage in India and is founder of the Yorkshire Jain Foundation in Leeds, England. He is also a Professor of Statistics with an academic background in Mathematics and Physics, through which he has reaffirmed his spiritual heritage. Aidan Rankin, by contrast, is not a Jain himself, although he has written extensively on Jain thought. He is a London-based writer with a social science background, including a PhD in Government from the London School of Economics. However, he realized that the social sciences are merely scratching at the surface when they ignore the human spirit and focus exclusively on statistics and methodology. For both writers, Jain science is a humanistic discipline in the true sense, balancing reasoned inquiry with compassionate intuition.

In this sense, the book is simultaneously a meeting of minds, an encounter between disciplines and a synthesis of our earlier research, producing something new. Kanti Mardia's book, *The Scientific Foundations of Jainism*, was published in Delhi by Motilal Banarsidass in 1990 and broke new ground in connecting the sum of the most ancient philosophical speculations of the Jains with the terminology of modern physics and logic. In *Living Jainism*, he develops this theme more deeply and applies it to new areas of scientific and social interest. Aidan Rankin's books, *The Jain Path: Ancient Wisdom for the West* (2006) and *Many-Sided Wisdom: A New Politics of the Spirit* (2010) sought to apply Jain principles to some of the most pressing cultural and ecological questions affecting the present phase of human development. In this book, his journey continues and his search for a new way of thinking about social and political questions mirrors his co-author's speculations about the physical sciences.

The academic study of Jainism has gathered momentum in the West over the twentieth and early twenty-first centuries, as the opportunities for cultural encounter have widened and the old certainties of economic and cultural imperium have been steadily eroded. With the emergence of India as a global power – through both the 'hard' power of economics and the 'soft' power of cultural influence – interest in ideas of Indic origin has correspondingly increased. Although Jain studies still remain a specialism confined to a relatively small number of experts, there is a growing sense that the ancient yet highly evolved teachings of the Jains have an extraordinary relevance for our time. Jainism, it might be said, is 'in the air': there is an inchoate awareness of its potential to enlighten and enrich us.

Scholars such as Professor Padmanabh S. Jaini of the University of California, Berkeley, have played a crucial role in nurturing this growing consciousness of the Jain tradition. His seminal work, *The Jaina Path of Purification*, has become the

starting point for all manner of explorations of Jain culture, history and ideas. Without this great *tour de force*, a comprehensive survey of Jain philosophy and practice, as well as the cultural and historical conditions underlying them, it is likely that Jainism would remain little understood by western academia and be an object only of brief overview or obscure speculation. Certainly Professor Jaini's work is the primal inspiration for our study, both as an inspiration and a principal point of reference, and we pay personal and intellectual tribute to his achievement.

## Chapter Outline

In the following ten chapters (which include the Epilogue), we seek to paint a picture of Jain philosophy with a focus on its relationship to modern science. Our book is aimed at the interested inquirer of any background and is as accessible to the non-specialist at least as much as to the student or academic.

Chapter 1 lays the foundations of our discussion by explaining how Jain culture – which we call *Jainness* – subtly guides its adherents, and gives some impression of what it means to be a Jain, in terms of both thought and practice. The twin ideas of respect for all living beings and all life as interconnected are introduced as core teachings. The chapter examines the history of Jain teachings and distills them into Four Axioms, or 'Noble Truths', that closely link them to scientific terminology. The Noble Truths together form the backbone of our study.

Chapter 2 unveils one of Jainism's treasures, a system of logic that enables all shades of opinion and meaning to be assessed before any judgment is passed. The process associated with Jain logic is a form of mental meditation, but it is also a manifestation of intellectual Ahimsa. In Jain teachings, it must always be remembered that thoughts *are* actions. The underlying thought or intention reveals the true character and informs its consequences. Jain logic is also a scientific method that clarifies thinking and enhances the sense of multiple possibilities that the researcher should keep in mind. Many-Sidedness is an aspect of Jain logic that can have intense ramifications for spiritual practice, politics and society as well as 'pure' science.

Chapter 3 explores the distinctive Jain conception of the soul or Jiva (a term used frequently throughout the book). It introduces the conception of karma as an expression of subtle matter, an idea that contrasts markedly with other Indic traditions. The nine levels of reality experienced within Jain teachings are also explained and given a wider scientific context.

In Chapter 4, we revisit the key teaching of respect for all forms of life as an aspect of Ahimsa or non-injury. Jain theory is based on an interdependent 'hierarchy of life' based on mutual responsibility. Levels of consciousness, including spiritual consciousness, differ between and among species, with a higher level of responsibility corresponding to higher levels of consciousness.[3] Humans are capable of the highest forms of intellectual and spiritual development, but also of the spiritually lowest and most destructive behaviors. The hierarchy of life stretches beyond the earthly or material realm to include beings and forms not directly perceptible to humankind. The four states of existence (or 'destinies') are thereby introduced along with the Jain use of the *svastika* as a symbol of life energy – in contrast to the deathly distortion of this image in twentieth-century Europe.

Chapter 5 continues with the theme of universal cycles in Jainism, as they also relate to the cycle of birth, death and rebirth for the individual soul. Types of karmic influence are explored in greater detail, as are the concepts of space, time and matter in Jain teachings.

In Chapter 6, we turn towards the psychological effects of karmic influences, notably the *Kashaya* or passions which block the pursuit of knowledge. They do this by generating delusions of power and accentuating irrelevant and harmful desires. These can have dangerous consequences for society and the environment, but above all they adversely affect the individual's inner life.

Chapter 7 posits more concrete definitions of Ahimsa and its negative counterpart, *Himsa,* and their implications for spiritual and scientific practitioners alike. The principles of 'positive non-violence', Careful Action and the reduction of harm to a minimum are clarified as guides to the conduct of life and thought.

In Chapter 8, the psychology and ecology of Jain teachings remains the focus as we explore the process of *purification*. From a Jain standpoint, this means clearing the mind of passion and dogma (something that any good scientist should do!) and moving away from some of the most cherished human delusions, in particular ideas of dominating and exploiting nature rather than learning to work with it. Knowledge increases through open inquiry and also through living more simply and being aware of our earthly limitations as well as the possibilities conferred by our intelligence.

Chapter 9 takes this theme further and explores the concept of Austerities in Jain culture, as a means of reducing the karmic influences that clutter our minds and inhibit the transmission of knowledge. Austerities have another powerful resonance, for they represent practical ways to reduce consumption and rid ourselves of its adverse psychological consequences. Once again, austerity does not mean joyless Puritanism, but the liberating experience of learning to tread more lightly.

In the Epilogue, we conclude by examining the direct relationship between Jain philosophy and recent developments in scientific thought. The works of Albert Einstein and David Bohm, along with the emergence of particle and quantum physics have led to a move away from linear towards holistic thinking. Increased environmental awareness has led, in turn, to a sense that we should rebalance our relationship to 'the rest' of nature, remembering that resources (including our own inner resources) are finite.

Appendix 1 gives a brief life of Mahavira, and Appendix 2 compares and contrasts Jain concepts with some other viewpoints.

Throughout these chapters, we have looked at Jain philosophy and culture in terms of their relevance to universal human problems. Although we frequently hold up Jain teachings as a critical mirror to western science – including the social sciences – we do not assert its superiority to other systems of thought. Yet we *do* argue, we hope persuasively at times, for Jainism's special qualities and relevance to the current human predicament. We also maintain that Jain science provides an effective mental training for those who wish to address scientific or social questions. In a world that is increasingly connected at some levels, but radically divided at others, the practice of Careful Action can bring healing and the prospect of wholeness. The Jain perspective sheds a gentle light on these urgent problems, exercising a calming influence while pointing towards new ways to live and – just as importantly – new ways to think.

# Chapter 1

# Jains, Jainism and 'Jainness'

## Spiritual Victors

The first line of the most important Jain prayer is *Namo Arihantanam,* a simple phrase that means: 'I pay my profound respect to any living person who has conquered his or her inner enemies (or his or her lower nature).'

This is not a prayer in the sense in which the term is understood in most of the world's religious traditions. Instead, it is an invocation of human beings like us, past and present, who have attained the highest level of spiritual development through the use of both reason and intuition. At the same time, this prayer is a reminder of human possibility. Each one of us has the potential to conquer the self, to rise above our 'lower nature' and its illusions and arrive at our 'true nature' or inner self. Such possibilities are open to all, entirely irrespective of the religion, caste or social status of the individual.

The path of the Jain is based on reasoned thought and its practical application to everyday life. It is about knowledge and understanding rather than mere acceptance of inherited doctrines. Above all, Jainism is centered upon each individual's capacity to think, learn and discover, and then to apply that knowledge to personal conduct, priorities and values, and relationships with others (not just humans). Therefore Jain spiritual practice encompasses the disciplines of psychology and philosophy as well as the methods of the scientist. The prayer 'Namo Arihantanam' expresses a profound belief in human creativity, integrity and intelligence, reminding us that Jainism is a humanistic philosophy as well as a spiritual system that transcends ordinary human concerns.

*Jainism* itself is a term derived from the word *Jina* in the ancient Indian language of Arda-Magadhi. This language is contemporaneous with and related to Sanskrit and so was common parlance in parts of India some 2500 years ago. At the time, many of the outward forms of Jain belief practice familiar to us today were taking shape, although the tradition is far more ancient than that. The word *Jina* (or *Jin*) in Arda-Magadhi means 'the person who is a spiritual victor', in other words one who has conquered his or her self. Jainism is now taken to mean the religion followed by Jains, in other words those who seek to work towards self-conquest or spiritual victory.

The traditional greeting used by Jains is *Jai Jinendra,* which means 'Honor to the Supreme Jina'. This is a reverential acknowledgement of the success of those who have achieved a high level of scientific knowledge and spiritual understanding, and so inspire others to lead better lives. Equally, it is a 'democratic' acknowledgement that each of us carries the potential for self-conquest and the pursuit of knowledge. The dual implication of 'Jai Jinendra' tells us much about the nature of Jain teachings and what it is to be a Jain.

Much has already been written about the *ism* in Jainism, in other words the body of doctrines and practices of the Jains, the Jain ideology as it has evolved over many centuries and under many different political and social conditions. This book also contains many descriptions and analyses of Jain teachings. In particular, it looks at the ways in which Jain doctrines and insights often correspond – and at some other times contrast – with the secular approaches of modern sciences such as physics and cosmology. However, rather than taking as our starting point the systematized ideology – the ism – of the Jains, we have chosen to focus on the sensibility, the way of looking at the world that goes with being a Jain.

We have chosen this approach because we believe that the Jain *way of thinking* has distinctive qualities with significance beyond Jainism as a cultural tradition. The methods used by Jains to search for truth present a radical challenge to 'mainstream' patterns of thought today, as do their underlying attitudes towards knowledge, intuition, the nature of power and the place of humanity on Earth and within the cosmos. Simultaneously, the Jain approach to life accords with a remarkable number of the most pressing modern concerns, whether they are collective or individual, global or local.

In addressing the growing ecological crisis – and the intellectual as much as environmental challenge it presents – or in making sense of the human conflicts in our interconnected yet obstinately divided world, the Jain sensibility offers us guidance of subtle power and depth. For the individual seeker of knowledge, the Jain attitude towards truth can help negotiate a pathway among a bewildering array of competing viewpoints. In this way, pluralism can become a source of integration and strength, at both personal and social levels, rather than an agent of fragmentation and conflict as it is all too often today.

The Jain way of formulating ideas also sits well with the methods and practices of cutting-edge science. In their increased emphasis on the critical role of minute particles and micro-organisms, today's physicists and ecologists can be said to be catching up with centuries of Jain awareness. Concepts such as dark matter or even parallel universes have long been within the range of Jain consciousness. The sense that all life forms and all parts of the cosmos are linked and mutually dependent is the defining characteristic of the Jain world view. It is not so much an 'idea' as the intuitive leap from which the entire Jain quest for knowledge and spiritual attainment stems. In a world where materialistic notions of progress now seem uncertain or untenable, Jain thought offers a powerful critique and an alternative vision of progress. This has been arrived at through an alternative way of reasoning that matches the present complexities of science, society and the human psyche.

For Jains, the thought or intention behind an act can be at

least as important as the act itself. In the same way, the thought process from which Jain doctrines have arisen can be at least as important as those doctrines. For the purposes of this book, we shall understand this sensibility as 'Jain-ness'. Jainness gives rise to and evolves along with the body of knowledge, understanding and teachings known as Jain-ism. While many of our arguments and ideas are shaped by Jainism as a body of knowledge, our underlying wish is to convey an understanding of Jainness, the process of *thinking like a Jain,* because this could have a universal application and validity. Indeed it is possible that we are experiencing a 'Jain moment', as on a global level established 'truths' face unprecedented challenges and at the same time there is a frantic search for certainty and continuity. Parallel to this, a vast expansion of human knowledge is matched by a growing sense of what we do not know and the forces we cannot control.

Jainness is concerned with reconciling continuity and change, possibility and limitation. It offers the opportunity for spiritual liberation through self-knowledge, while accepting the mental and physical limits of the human form, the so-called 'mortal coil'. More than that, it provides a welcome and much-needed corrective to the dogmatic and absolutist systems of 'knowledge' (be they religious or secular, scientific or political) in which so many obvious cracks are appearing today. Rather than asserting or seeking to impose its own 'truth', Jainism asks us to look inside ourselves, find our own and continuously question it. Jainness is the disposition or frame of mind that enables us to begin this process.

## Origins of Jainism and Jainness

Loosely speaking, Jainism was founded by human spiritual guides known as *Tirthankaras.* This word means ford-makers, path-finders: those who point the way through or the way ahead. Tirthankaras are, therefore, the people who show us the true way across the troubled ocean of life, the leaders on a spiritual path.

They lead by example and wise counsel. They unlock knowledge by training our (and their own) minds to think in ways that point towards spiritual awareness and a sense of the real self.

In all, there were twenty-four Tirthankaras. The first of these was Rishava, an ancient lawgiver who, according to one Jain text, lived 'many thousands of years' ago. This implies that he lived before reliable recorded histories began and so he can be seen as a semi-mythological figure, somewhat like Lycurgus for the ancient Spartans. Rishava is recognized by some Hindus as a manifestation of the divinity Shiva – an illustration of the complex relationship between the Jain path and the Vedic or Hindu tradition, to which we shall later refer. It seems that Rishava as lawgiver was able to fashion out of several nomadic communities a settled, pacific society, based on the cultivation of grain, with cities, currency, trade and a common legal system. Those familiar with the Jewish or Christian traditions might find parallels with the figure of Moses.

Rishava appears eventually to have abandoned political power and material possessions in favor of the life of austerity and wandering. He came to identify material and worldly concerns with spiritual bondage, which Jains have come to think of as *karmic bondage*. In this, he set a precedent for other Tirthankaras and established the pattern of thought and the priorities familiar to Jains ever since. That is to say, he favored civilization, justice and knowledge over barbarism, chaos and ignorance. However, he also sought out a spiritual path beyond both the pre-conscious state of ignorance and the artifices of a sophisticated urban society. The former precluded enlightenment or rational thought. The latter provided insidious distractions and delusions of grandeur that became powerful obstacles to spiritual advancement.

This insight gives rise to a new interpretation of power, which is not based on controlling and subjugating others, acquiring material 'things' or claiming a monopoly of truth. In contrast to the conventional, outward vision of power, it is an inward vision, consciously repudiating the trappings of status and embracing humility. Yet it is also a more supple, flexible and enduring form of power that outlasts states, politicians and the wealthy. Such an interpretation of power matches the understanding of 'conquest' implicit in the term 'Spiritual Victor'. The struggle is for control of the territories within rather than geographical expansion or personal gain, as we usually understand conquest. These counter-cultural definitions of power and conquest are linked to an inclusive definition of wisdom as the open-minded pursuit of wisdom and insight into the workings of the universe and our place within it. Rational thought, meditation and the conquest of superficial desires can give us a sense of proportion. They bring us closer to an understanding of what is truly important in life, in place of the material cravings and ambitions that so easily bind us. This sense of equanimity is central to Jainness.

The spiritual discipline of Rishva was followed by twenty-one successive Tirthankaras until, in the age of the 23rd Tirthankara, Parshva, the tradition that we now call 'Jainism' began to appear in a form recognizable to us today. Parshva is also the first of the Tirthankaras whose historical existence as a human being can be verified reliably. He is generally held to have lived around 2800 years ago (traditionally dated 872-772 BCE). The philosophy and logic of modern Jainism emerged in a systematized form at the time of the 24th Tirthankara, Mahavira, who was born in 559 BCE and whose *nirvana* (full enlightenment) took place in 527 BCE. Mahavira means 'Great Hero', a name symbolic of spiritual victory through non-violence and the refusal of conventional power or wealth. He was a contemporary of Gautama Buddha (563-483 BCE), the overlap being thirty-six years, but there is no evidence that they ever met. It is also worth noting that the Buddha was in the process of enlightenment at the time when Mahavira was at the peak of his career (for a more detailed account of Mahavira's life, see Appendix 1).

Image 1: *Parshva, the 23rd Tirthankara (Digambara image: characteristically, the eyes, lips and torso are not marked). His idol is distinguished by the emblem of a snake. From Leeds, UK*

Even today, it is still not unusual for these two great spiritual teachers to be confused and conflated. It is sometimes claimed (against all available evidence) that Buddha and Mahavira were one and the same, or that Jainism is really a subset of Buddhism, which is far better known outside Indic civilization. There are many differences, both obvious and subtle, between the two philosophies, as well as areas of common ground, but such comparisons lie outside the scope of this study. In iconography, a simple distinction can be made by clothes: Mahavira is usually naked, whereas the Buddha is usually clothed! Icons of Parshva and other Tirthankaras are also frequently found in Jain homes and places of assembly.

To bring the dates of Mahavira and Gautama Buddha into a western perspective, we may note that Aristotle was born in 384 BCE and Jesus Christ around 4 BCE. India officially celebrated the 2500th anniversary of Mahavira's nirvana between 13th November 1974 and 4th November 1975. One of the strongest admirers of the Jain religion was Mahatma Gandhi, whose thoughts and actions were greatly influenced by certain Jains, in particular Raychandbhai (Raychandbhai Mehta, also known as Shrimad Rajchandra). Gandhi was inspired by Jain doctrines of non-violence, respect for life, ecological responsibility and the value of each individual. He applied these teachings to his strategy of *Satyagraha* ('truth-struggle') and non-violent resistance to British colonial rule. They also helped him shape his economic philosophy of *swadeshi:* self-sufficiency through co-operatives and local production for local need. While embracing many Jain ideas, Gandhi remained a devout Hindu and indeed credits his Jain guides with increasing his understanding of what Hinduism was really about. The evolution of Gandhi's thought is surely one of the clearest illustrations of how Jainness works.

Mahavira is often erroneously referred to as the 'founder' of Jainism. In reality, he sculpted a new form from material that had already been long in existence. Jainism has evolved out of the most ancient Indic spiritual teachings, which are in turn aspects of the earliest human consciousness of the universe. Living as a Jain implies awareness of *Dharma,* which is at once a natural law governing the workings of the universe and an ethical system showing us how we should live. Spiritual practice and the

Image 2: *Mahavira, the 24th Tirthankara (Svetambara image: characteristically, the eyes, lips and torso are marked). His idol is distinguished by the emblem of a lion. From Sirohi, Rajasthan*

pursuit of knowledge bring us into closer alignment with Dharma.

## Some Characteristics of Jainism

The most important principle of Jainism is that of non-violence in thought and deed towards fellow human beings and all other forms of life, including the smallest. Thus most Jains are vegetarians. Even honey and alcohol are avoided because they are believed to contain microscopic life. In the West, vegetarianism carries the connotation of a restrictive and puritanical diet, but in the case of the Jains, nothing can be further from the truth. Their food sources are carefully selected to minimize harm to living systems, but the results are rich and varied. In Gujarat, where Jain cultural influence is historically strong, the overwhelming majority of the population follow a vegetarian diet, noted for its gentleness and subtlety, and influenced by Jain tastes. This is another example of the 'soft power' of Jainness.

Jains therefore appreciate that even the smallest of organisms have life, including those that are invisible to the naked eye. Each life form has value and purpose, both as an individual and as a species or type. Each plays its part in the natural order and each is part of the cycle of spiritual (as well as physical) evolution. Jains recognize that even the most seemingly 'primitive' forms of life can profoundly affect nature's fragile equilibrium. Climate scientists today are becoming more aware of the importance of plankton in the ecology of the ocean and its consequent effects on global temperature patterns. This type of insight accords entirely with the Jain perspective.

Non-violence and respect for life in its infinite variety form the core of Jain doctrine. All other teachings and practices derive from these simple principles and their radical consequences for human life and thought. It becomes clear, from this standpoint, that all living systems are interdependent: they literally *need* each other. As living beings, 'we are all in this together' as part of a process that encompasses spiritual development and environmental equilibrium. This extends to micro-organisms, plants and non-human species. Moreover, Jainism has always fully acknowledged the possibility of other forms of life in other areas of the universe. Jainness, therefore, is an inclusive rather than *anthropocentric* (narrowly human-centered) state of mind. This approach dovetails neatly with both scientific ecology and the accompanying social and political movements. It can justifiably claim to be one of the oldest or most original expressions of consciousness. Truthfulness, refraining from stealing or economic exploitation, moderation in acquiring personal possessions and avoidance of sexual promiscuity or exploitative relationships are other important facets of Jain teaching. They arise from the underlying principles of non-violence, sacredness of life and the need to treat others with dignity and respect. Meditation, general self-control and the search for a state of equanimity are fundamental to the Jain outlook and way of life. They make it possible to live harmoniously with nature, with others and ultimately with oneself.

Jains do not believe in any external God who created and sustains the world, and neither do they believe in any means of redemption outside of themselves. The individual has to achieve his or her own salvation through a combination of Right Faith, Right Knowledge and Right Conduct, known as the Three Jewels or *Tri-Ratna*. Salvation is believed to terminate the cycle of births, deaths and reincarnation (*samsara*). Crucially, the liberation of the soul is equated with infinite knowledge as well as eternal bliss, in other words freedom from intellectual constraint as well as material and physical bondage. This gives knowledge and learning a sacred status and a central place in Jain culture, which has always laid great stress on the education of men and women of all social classes.

Spiritual liberation is also equated with release from karma. We shall explore the Jain theory of karma in some detail in ensuing chapters because it is such an essential ingredient of Jain science. Karma in Jainism differs radically from its Hindu or Buddhist counterparts. Rather than an abstract cosmic law, it is seen as a substance, a form of subtle matter that weighs down and encases the soul. By so doing, it obscures true knowledge and understanding of reality. Involvement with karma keeps the soul imprisoned in the cycle of samsara and hence keeps the individual preoccupied with material attachments that prevent a clear understanding of reality. These attachments, in turn, generate more karma and so block out knowledge, much as a build-up of dust on a window pane can hide the light. The task of the spiritually conscious individual is to reduce such negative patterns of attachment. Through this process, true priorities are recognized, hence the inflow and accumulation of karma is reduced.

From this description, we may infer that most if not all responsibilities are devolved to the individual. This is certainly true in the sense that because every life is sacred and precious, individual liberty and freedom of conscience are sacrosanct.

Jains have a profound belief in human (or sometimes non-human) intelligence and individual initiative is highly prized. Probably, this is one of the reasons for the remarkable success of many Jains in the world of business. Such initiative is equally applied to spiritual as commercial or professional advancement! That said, it would be quite wrong to view Jainism as individualistic in the narrow sense of caring only for the self. On the contrary, seeking material (or any other) advantages for oneself at the expense of others is regarded as the most spiritually dangerous behavior. It follows that a self-centered mentality – and the actions that arise from it – will generate the heaviest and most negative levels of karma, and point in the opposite direction to self-knowledge. A key component of understanding one's self is understanding the equal worth and value of other 'selves'. Thus our individuality is realized by co-operating with and caring for others and minimizing harm to all life forms. On the 2500$^{th}$ anniversary of Mahavira's nirvana, the Jain community adopted as a motto *Parasparopagraho Jivanam,* meaning 'Souls render service to one another' or 'All life is bound together by mutual support and interdependence.' This phrase is combined with an emblem depicting an open hand that contains in its palm the word Ahimsa, or non-violence, representing the first principle of Jain thought.

In practical terms, this means that Jains are encouraged to make the best use of their abilities *and* use them as far as possible for the benefit of society. This is why so many Jains may be found working in education, medicine or scientific research. There is also a very strong tradition of philanthropy, charitable giving and voluntary work. In India, Jain businesspeople are noted for the number of schools, colleges and free hospitals they have founded and the scholarships they have endowed for poorer students. They have founded and directed shelters for the homeless and animal refuges, such as the internationally renowned sanctuaries of Ahmedabad. These two aspects of Jainness, individual autonomy and consideration for others, do not cancel each other out. Self-help is considered a part of the spiritual journey, while the commercial success of many Jains is founded on long-term thinking and careful action. This latter principle influences all aspects of life and is kept constantly in mind. Jains, therefore, are not asked to make artificial 'choices' between individual freedom and social responsibility. Among Jains, there is no single leader such as a 'Pope', neither does any person have supreme authority. However, there are monks who have chosen to devote themselves entirely to spiritual matters and who lead very simple lives with minimal attachments. Certain teachers and lay leaders, past and present, are given particular respect, but this respect has had to be earned through scholarship and service. There are many scriptures (see below), but no definitive text equivalent to the Bible, Torah or Koran.

Umasvati's *Tattvartha Sutra* (That Which Is), written in the second century CE, is the most comprehensive single treaty on Jain teachings. Nonetheless, the onus is on the individual to find the truth for himself, through meditation, rational thought and practical experience. In short, each individual is his or her own guru. The principles of Jainism are intended to be self-verifying so that the seeker discovers truths for him- or herself, rather like a research worker in a laboratory.

Jainism is a scientific philosophy in the literal sense: the English word 'science' is derived from the Latin verb *scire,* to know. Knowledge is increased through critical examination, trial and error and the evolution of ideas. Therefore, all opinions, even those that appear to be manifestly 'wrong', should be respected and at the same time engaged with and challenged. This is not a standpoint of extreme relativism in which there is no objective truth. Right (and wrong) answers do exist but the former cannot be imposed from without. They have to be learned, experienced and tested. Truth exists, but has many angles and aspects, for which all of us should search with humility. The more we assert that we have found the 'whole truth', the farther we are likely to be from it.

Humility is a defining characteristic of the Jain mentality. It acts as an inoculation against ideological dogma, intolerance and rigidity, which give rise in turn to violent actions and thoughts. The intelligent seeker of knowledge needs to be sensitive to the limits of that knowledge: the more he knows the more he realizes that he has much to learn. Increasing our understanding therefore involves listening to alternative viewpoints and continuously questioning our own. It means being open to new aspects of the truth or alternative viewpoints to which we have previously been blind. We should be aware that the reasons *why* we believe something are as important as the truth of that belief, just as intention is as important as action.

This way of approaching the problem of knowledge derives

from the emphasis on non-violence and interconnectedness. Jainism does not accept that there is a division between thought and action. Thoughts are, in themselves, actions that have consequences, including violence and harm to others as well as creativity and insight. The principle of careful action therefore extends to the way we formulate our thoughts. In so doing, we are reminded to be aware of the desires – conscious and unconscious – underlying them. Such awareness makes us realize that our thoughts and beliefs are not 100 per cent foolproof and can always be examined and revised. Jainism's holistic view of nature, in turn, gives rise to a holistic view of knowledge. If all living systems are interconnected, then we should see ideas also as parts of a whole, rather than disconnected, abstract concepts. This holistic logic is known as *Anekantavada:* 'many-sidedness', or 'non-one-sidedness'. We shall explore that concept and its implications in more detail later in this study. For now, it will suffice to say that the Jain outlook sees beyond the adversarial, confrontational approaches that characterize so much of modern political, philosophical and scientific thought.

There are several differing schools within the Jain tradition. The two main ones are *Digambara* ('Sky-Clad') and *Svetambara* ('White-Clad'). These labels are based on the attitude of their ascetic orders towards clothing. Whereas Svetambara monks and nuns wear simple white garb, the Digambara male ascetics are naked ('sky-clad'). Digambaras believe that their monks should renounce everything in the search for full enlightenment. This is because the Jina can manifest no worldly activity and has no bodily functions. Svetambaras emphasize simplicity and modesty in living; they advocate this rather than total renunciation.

These two schools both have extensive lay and ascetic followers. Both seek to live by the principles of *Ahimsa* (non-violence, avoidance of harm) and the ethics of restraint and compassion that arise from it. Ascetics, however, submit to a far stricter self-discipline than their lay counterparts, following a stricter, more literal version of the Five Vows (*Vrata*):

- *Ahimsa:* Non-violence, non-injury, respect for all life
- *Satya:* Truthfulness, honesty, personal integrity (including avoidance of hypocrisy)
- *Asteya:* Avoidance of theft, taking what is not given, and exploitation
- *Brahmacharya:* Chastity, avoidance of promiscuity and exploitative relationships
- *Aparigraha:* Non-possessiveness, avoidance of wasteful consumption or unnecessary accumulation of possessions

Ascetics practice the *Mahavratas* (Greater Vows) and lay men and women the *Anuvratas* (Lesser Vows). A good example of this is Brahmacharya, which monks and nuns interpret as meaning a vow of celibacy, whereas lay men and women regard it as meaning fidelity in relationships. The popular image of the Jain monk sweeping the ground before him lightly with a brush stems from a literal interpretation of *Ahimsa*, by which the ascetic must endeavor to avoid harming even the tiniest forms of life. This gesture also powerfully symbolizes the careful action and measured thought to which all Jains aspire. Svetambara and Digambara Jains use 'idols' or 'icons' as objects of contemplation. These are usually small statues hewn from metal or stone, either of specific Tirthankaras or generic images of the Jina. These are meditative tools that inspire Jains and remind them of the principles they are attempting to live by from day to day. They act as a focus for calm, rational thought and symbols of Jain culture. There are a few identifiable differences between the idols of the two schools. In Svetambara idols, for instance, the eyes, lips and torso are marked, unlike those of their Digambara counterparts (see Images 1 and 2 above). There have also been various reform movements through the ages. Two sub-groups of

Svetambara, called Sthanakavasi and Terapantha, do not believe in temples and reject the use of idols. The traditional place of worship and social center for Jains is the temple or *Mandir*. A sub-group of Digambara, called Taranapantha, has also proscribed the use of idols. Despite their different emphases, the essential tenets of Jainism are followed by all schools and sub-groups; this includes belief in the Twenty-Four Tirthankaras as supreme exemplars. All practicing Jains aim for the ultimate goal of *Moksha:* spiritual liberation through enlightenment.

## Jain Scriptures

It is believed that the sermons of a Tirthankara take the form of divine language or sound: divine in the Jain context means super-human, or beyond the conventional level of consciousness. According to the Digambaras, this sound transmits the intrinsic meaning of the teaching that is then translated into the scriptures by several chief disciples, 'ganadhara'. For the Svetambaras, by contrast, the Tirthankara speaks in a divine human language. In general, the role of the ganadhara was one of translator and editor. Thus, one should not take the scripture literally, but keep the idea of self-analysis and synthesis at the forefront of one's mind.

### 1. The Main Scriptures

In all there are sixty Jain scriptures (*Agamas*), which are classified into three parts:

1. Purva (Old Text)
2. Anga (limb)
3. Angabahya (Subsidiary Canons)

The Purvas are extinct, but the spirit of these ancient texts lives on and is reflected throughout the rest of the Jain canon. Gautama and Sudharman, Mahavira's chief disciples, have been the main contributors to the main twelve scriptures (*Angas*) but the tradition of oral transmission was carried on for a long time.

## 2. Secondary Scriptures

The secondary scriptures (*Anuyogas*) supplement the older material described above and there are four parts just as if they were the Jain equivalent of the four Hindu *Vedas*. The Anuyogas were mostly written by monk scholars:

1. *Prathmanuyoga* (the primary exposition) deals with biographies of Tirthankaras.
2. *Karananuyoga* (exposition on technical matters) deals with ancient sciences such as cosmology and astrology.
3. *Charananuyoga* (exposition on discipline) is the most important work on Jain Yogas. It includes Hemacandra's *Yogashastra* (twelfth century) and Haribhadra's *Dharmabindu* (eighth century).
4. *Dravanuyoga* (exposition on existents) includes the most important work, *Tattvartha-sutra* ('That Which Is'-verses) of Umasvati (second century CE). This work summarized concisely the whole of the Jain doctrinal system into about 350 verses. It is comparable to Patanjali's *Yoga-sutra* in that it presents the teaching in an integrated philosophical school.

Other works included are Siddhasena Divakara's *Nyayavatara* and *Sanmati-sutra* (fifth century CE), which are excellent works of logic. Yasovijaya (eighteenth century CE) represents the modern school of logic.

Our description above has been somewhat restricted to the Svetambara canon. The Digambara Jain tradition also holds that there were sixty texts with the above titles but believes that they are all lost. They possess some record leading to two important scriptures of the second century: *Satkhandagama* (scriptures in six parts), and *Kashayaprabhrta* (Four-Passions-'Gifts'). The work of Kundakunda (perhaps second century CE) is the most comprehensive and includes *Samayasara, Niyamasara* and *Pravacanasara*. His tradition was continued in the sixth century by Pujyapada. The important commentary *Atmakhyati* on Samayasara by Amrtacandra appeared in the eleventh century. Other representative writers to be mentioned are Jinasena (ninth century) and Somadeva (tenth century). Appropriate versions of Umasvati's work *Tattvartha-sutra* (as well as Siddhasena's work on logic) are accepted by both sects.

The first group of *Agama* scriptures was written in Ardha Magadhi, a Prakrit dialect originating in the Magadha region of modern Bihar. Prakrit was a vernacular or 'popular' alternative to Sanskrit, which evolved into a literary language of equal distinction, but remained unrecognized by the 'orthodox' Hindu Brahmins. Because of this, it became associated with the writing and dissemination of (in Brahmin terms) heterodox schools of thought. There were numerous versions of Prakrit, but the Maghadi version became the most widely used and is the most closely associated with Jainism: so much so, indeed, that the terms 'Prakrit' and 'Ardha Magadhi' are often used interchangeably.

Many subsequent Jain works are in Sanskrit, however, starting from the work of Umasvati. Thus, there is a vast literature available but it seems that *Tattvartha-sutra* of Umasvati can be regarded as the main philosophical text of the religion and is recognized as authoritative by all Jains.

## The Four Noble Truths

Every spiritual path starts from some form of conviction or belief system that can be learned and applied. In this book, we follow the teachings of the Jains condensed into four essential axioms by Mardia (1990), through which the whole path can be understood. These try to answer questions such as 'Why are we imperfect?' and 'What shall we do about it?' If we were all immortal, perfect, all-knowing and eternally happy, then there would be no place for any form of spiritual path. However, life's journey consists of 'many ups and downs', with pleasure and pain as the central themes of existence. The purpose of a spiritual path is to make sense of these fluctuations and to guide us through them.

Furthermore, one comes across all types of living entities who have widely varying life experiences and reactions. Spiritual practitioners ask questions such as 'Why do these differences exist?', 'Why are some born with disabilities and others without?', 'Why do some have special gifts and insights?', 'What is "good" and "bad"?', 'Is a "perfect" society possible?', 'Why are there different forms of life and how do we respond to these differences?' Jains take a long view of existence because of the belief that the soul journeys through many material existences and incarnations before the moment of liberation. This can include many human types and many animal or plant species, as well as entities popularly regarded as supernatural and (conceivably) extra-terrestrial beings. Therefore, the Jain mentality adopts an inclusive view of life and views biological and spiritual evolution as parts of the same process.

The four axioms of Mardia (1990) attempt to distill those aspects of Jain teachings that explore the *science of life*. They approach that science from both the physical and ethical standpoints, since Dharma is at once a physical and moral universe. These axioms have been called the Four Noble Truths, following Mardia (2008). This style of construction has been made widely familiar by the rapid spread of Buddhist teachings. The truths of Jain Dharma are compatible with, but radically distinct from, those of Gautama Buddha. In the spirit of many-sidedness, they approach reality from the cultural perspective of the Jains but have a universal resonance. Truth 4 is divided into three sections (4A-C). All four propositions examine the role of karma in the spiritual journey of the soul and the physical laws of the universe. Karma, which we shall frequently refer to in terms of 'karmic matter', 'karmic particles' and 'karmons', is fundamental to Jainism's scientific approach to the workings of the universe and human psychology.

Exploring these four axioms in detail gives us a clearer understanding both of Jain ethics and the processes and dilemmas of modern, secular science. The differences between the two approaches will become readily apparent, as will the often remarkable levels of correspondence. Our task is to demonstrate that there is a distinctive and living Jain Science that has relevance and value in its own right. In so doing, we wish to increase the level of interest in an ancient (yet also profoundly modern) faith. At the same time, we hope to contribute to the growing convergence between spirituality and science.

The Four Noble Truths are as follows:

Truth 1: *'The soul exists in contamination with karmic matter and it longs to be purified.'*

Truth 2: *'Living beings differ due to the varying density and types of karmic matter.'*

Truth 3: *'The karmic bondage leads the soul through the four possible states of existence (cycles).'*

Truth 4A: *'Karmic fusion is due to perverted or distorted views, non-restraint, carelessness, passions and activities.'*

Truth 4B: *'Violence to oneself and others results in the formation of the heaviest new karmic matter, whereas helping others towards Moksha with positive non-violence results in the lightest new karmic matter.'*

Truth 4C: *'Austerity forms the karmic shield against new karmons as well as setting the decaying process in the old karmic matter.'*

These axioms look directly at the roots of the tree rather than its branches. Their meaning and plausibility is discussed in relation to Jain scriptures, teachings and practice in the following chapters. Truths 1-3 postulate Jainism's scientific theory of karmons, whereas Truth 4A-C postulates the practical applications of this theory.

The meaning and significance of these formulae will become clearer in the chapters below, where we explore each Truth in turn and give it a wider context. Taken together, the four axioms are an expression of Jainism's scientific spirit and the rational, reflective disposition of 'Jainness'.

## Jain Timeline: Some important dates in Jain history (in italics) together with other key dates

*??? BCE: Rishava – 1st Tirthankara*

... 21 other Tirthankaras: a continuity of the Jain tradition
*872 BCE: Birth of Parshva – 23rd Tirthankara*

*772 BCE: Nirvana of Parshva*

*599 BCE: Birth of Mahavira – 24th Tirthankara*

563 BCE: Birth of Buddha

*527 BCE: Nirvana of Mahavira*

483 BCE: Nirvana of Buddha

384 BCE: Birth of Aristotle

? c.4 BCE: Birth of Jesus Christ

1869 CE: Birth of Mahatma Gandhi

1948 CE: Death of Mahatma Gandhi
*13 November 1974 CE: 2500th Anniversary of Nirvana of Mahavira*
N.B. Many of these dates (for example those of Parshva) are based on traditional sources and accounts

# Chapter 2

# Jain Logic

*There are more things in heaven and earth, Horatio*
*Than are dreamt of in your philosophy*
William Shakespeare, *Hamlet*, Act 1, scene 5, ll.165-7

## The Principle of Many-Sidedness

In the chapters that follow, we shall explore in some detail the Jain belief in the Jiva (otherwise known as the 'soul' or unit of life) and its journey towards purity, which is identified with omniscience and freedom from the karmic cycle. The purer the Jiva, the higher its levels of knowledge, perception, bliss and energy will become. Purity in this context means absence of karmic influence, which traps the soul in ignorance and false consciousness. It means freedom from passions (Kashaya) and ultimately all worldly activities (yoga).

Spiritual practice, like scientific research or the pursuit of any form of knowledge (*jnana*), involves delving beneath what is immediately apparent to discover concealed truths. It also requires a change in values, adopting a longer-term or more 'universal' perspective beyond small-scale, materialistic concerns. From a Jain standpoint, the pursuit of scientific and spiritual knowledge (two sides of the same coin) involves adopting a more *holistic* view of the self. This means first seeing past one's present embodiment or transient individuality. In this way, we learn to view the self as an evolving entity passing through stages of consciousness and outward forms. Secondly, we learn to be aware of our individual imperfections, and a wider human imperfection, in our search for knowledge.

Therefore, we pursue the truth with humility, because we have not yet achieved omniscience. We realize that truth is a *many-sided* phenomenon. It has multiple facets, not all of which we can grasp. It follows that viewpoints or approaches to the truth that seem unfamiliar, unsympathetic or even downright tendentious might contain elements of wisdom. Each of us could be said to be on the same journey towards the truth, whether we realize it or not. We perceive things differently, according to our experience and level of understanding, our cultural background and indeed our species. Every viewpoint is part of the spiritual journey of all souls and is worthy of consideration. By the same token, no single viewpoint has absolute credibility: it represents a strand or aspect of a larger whole.

The holistic principle of Jain logic is known as *Anekantavada* or often *Anekant*. Anekant is one of the central tenets of Jain philosophy. The word can be translated as Many-Sidedness, (acceptance of) Multiple Viewpoints or more literally 'Non-One-Sidedness'. *Ekant*, by contrast, is the quality of One-Sidedness or doctrinaire, rigid thinking, which is viewed as a barrier to the perception of truth.

Jains believe that only omniscience, the product of Moksha, can convey an entire understanding of reality and its aspects as enumerated in the 'Nine Reals' or Tattva. We are subsequently left with another four 'types' of knowledge: mind and senses; Jain scriptures and the two faculties of clairvoyance and mind-reading. The last two are not widely esteemed in the West, where they are associated with magic and pseudo-science. However, they are revered within Indic civilization as a way of attuning the mind to other dimensions. Jains are ready to admit such possibilities because they are aware of how much we, individually and collectively, *do not know* about the universe and the powers inherent in the natural world.

In other words, the Jain position differs equally from the naïve openness (some would say gullibility) associated with large areas of New Age thinking and its adversarial counterpart, hyper-rationalism. This demands extreme scepticism about any spiritual dimension to the universe, or indeed anything that cannot be conventionally measured or quantified. The Jain mentality is eternally curious and open to new ideas and possibilities in the pursuit of knowledge. At the same time, it is understood that no new idea or fact can represent the 'whole truth' and nor can established 'facts' be allowed to congeal into dogmatic abstractions.

The cultural disposition we have defined as Jainness adopts a pluralist view of the truth – but one very different from the post-modern relativism that has become fashionable as a reaction against Enlightenment certainties. At its most extreme form, such relativism argues that 'objective' truth does not exist and there are only competing subjectivities. From the Jain standpoint, by contrast, truth exists but it is made up of many parts and can be viewed from many angles. This approach accords well with particle physics, which is based on continuously revising our understanding of reality: not repudiating previously established truths, but either enlarging upon or refining them. The many-sided nature of truth gives it its strength and indeed its sacred character. It is identified with Dharma, which is reflected in the complex workings of the universe.

There is much of value and interest to the modern scientific thinker in the Jain theory of knowledge. Broadly speaking, the theory has three major components:

1. *Pramana* (organs of knowledge / the approved means of knowledge)
2. *Naya* (standpoints / philosophical standpoints)
3. *Anekantavada* (holistic principle of Many-Sidedness), of which *Syadvada* (conditional predication) forms an essential part.

Knowledge about an object is obtained through two processes: a partial process and a total process. The total process is termed *Pramana,* literally meaning 'organs of knowledge'. This method gives credence not only to observational processes but also to the process of mental faculty. Because of the involvement of both these faculties, it gives a complete view of an object. The partial process is called *Naya* (standpoint). This involves the study of an object with respect to one single aspect at a time. There may be many aspects and so there could be many standpoints. This, however, does not give a complete picture of the object.

There are two types of organs of knowledge: direct and indirect. The direct has two varieties: sensory and super-sensory. Sensory knowledge consists of processes of recollection, recognition, concomitance and *inference*. The syllogism discussed below is an example of the use of inference. Super-sensory knowledge is the type of understanding associated with the 'paranormal' in western culture: *Avadhijnana* or clairvoyance; *Manahparyayajnana* or mind-reading and *Kevalajnana* or infinite knowledge. The latter quality is available only to the fully liberated soul (i.e. the complete individual consciousness), while the former are accepted as probably being outside the reach of most spiritual practitioners and only available to a few adepts, with continuous effort and training of the mind. Nonetheless, such qualities of mind may be aspired to by all (in this or future incarnations) and the practice of meditation by lay men and women can lead to an altered or more receptive state of consciousness.

The presence of the paranormal within a system of humanistic logic might seem surprising to the modern student, but there are two closely overlapping reasons why it is there. First, it is a survival of the most ancient forms of Indic spirituality, when there was no separation between the quest for knowledge and a sense of awe and reverence for nature. Intuitive thinking predominated and with it came a sense that there are untapped powers latent in the universe and within each of us. Primal intuition is one of the foundations of the Jain tradition, onto which layers of accumulated wisdom and rational insight have been grafted over the centuries.

Intuitive power is also regarded as a valuable counter-balance to reason, which can solidify into dogma if unchecked. The synthesis of reason and intuition provides the second reason why paranormal qualities can fall within the framework of Jain logic. Reason and intuition both point towards the idea – alluded to above – that there is much still to be discovered. This idea is also readily understood in western thought (see the Shakespearean epigraph to this chapter) and is arguably the founding principle of the discipline of physics.

The *Naya* (standpoints) serve as a base for the full comprehension of an object through gradual stages. Comprehension can come about in two ways: on the basis of the properties of the object and on the basis of verbal expressions about them. The Naya therefore start from an overall picture of the object, then break it down into its component parts, which consist of innate qualities *and* the way the object is perceived and described by observers. Then, these components are brought together into a unified form. The process of working through the Naya to define an object – or an idea – may be likened to the progress of the Jiva itself, beginning in a pure but naïve form then working through various stages of incomplete consciousness before achieving wholeness and enlightenment.

We shall now briefly examine three specific processes of Jain logic: the syllogism, the conditional predication principle and the holistic principle.

## The Jain Syllogism

A typical Jain syllogism consists of five propositions, for example:

1. Tom died, Dick died and so did Harry.
2. Tom, Dick and Harry are truly universal types of men.
3. Therefore, all men die.

4. John is a man.
5. Therefore, John will die.

The last three terms of the medium syllogism can, of course, be recognized as the Aristotelian syllogism which would be:

1. Man is mortal.
2. John is a man.
3. Therefore, John is mortal.

The syllogism clearly combines inductive and deductive methods of reasoning. In fact, it reflects the main stages of scientific/statistical thinking. The first two terms can be thought of as taking observations from a population and the third term as drawing inference from the observations. The last two terms give a projection about a new observation. This empirical logic is the basis of scientific methods and should not be lost sight of in all scientific applications.

In fact, the syllogism is said to be accurate when all its five parts are in harmony with each other. For the example above, the five parts are:

1. John will die
2. Because he is a man
3. Like Tom, Dick and Harry
4. As they died
5. So he will die

A syllogism is said to constitute a fallacy (*abhasa*) if any of these five parts are discordant with our observations.

## *Syadvada*: The Conditional Predication Principle

A central feature of Jain logic is the principle of conditional predications known as *Syadvada*. Inferences are examined from seven standpoints (*Saptabhangi-naya*), prefixed by the idea of 'maybe':

1. Maybe it is (from one standpoint) = *syadasti*
2. Maybe it is not = *syatnasti*
3. Maybe it is and is not = *syadasti nasti ca*
4. Maybe it is indeterminate = *syadavaktavyah*
5. Maybe it is and is indeterminate = *syadasti ca avaktavyasca*
6. Maybe it is not and is indeterminate = *syatnasti ca avaktavyasca*
7. Maybe it is, is not and is indeterminate = *syadasti nasti ca avaktavyasca*

(Maybe = *syat* or *syad*. Indeterminate = *avaktavya*)

Note that all predications have a margin of uncertainty, and each of the seven predictions may be called a Naya as it represents one aspect of an object. Predication (1) can be visualized as 'green' at a set of traffic lights, (2) as 'red'. Its special feature is (4) which allows for the possibility of indeterminacy, i.e. 'amber'. Other predications are syntheses of (1) and (2) with (4).

Syadvada is sometimes loosely translated as 'maybe-ism'. This conveys the idea of uncertainty, or thinking *around* a problem rather than cutting swiftly towards an 'answer' that is likely to be one-sided or partial. However, 'maybe' is not ultimately the best rendering of the word 'syat', which is more accurately translated as 'from one standpoint'. For Syadvada is anything but indecisive. It is a process of working through a range of partial explanations to arrive at a holistic solution.

The Syadvada principle has been formulated in terms of Boolean Algebra by Professor G.N. Ramchandran (1982). He applied the idea of Syadvada to mathematical logic, introducing the concept of multiple possibilities.

## *Anekantavada*: The Conditional Holistic Principle

We have just described the Jain method of looking at sub-parts of a problem through conditional predication, in other words exploring a range of partial solutions and using them to construct something nearer to the truth. One aspect of Syadvada is that it is an infinitely flexible system, absorbing new possibilities as they arise or are discovered. Syadvada is itself an aspect of a larger system, *Anekantavada* (*Anekant*) or Many-Sidedness. This is both a way of looking at the world – a *Weltanschauung* – and a holistic principle used to formulate logic. Anekant combines aspects of knowledge into a whole through a more complex version of the syllogism we considered above.

Consider first the following example. There are six blind men who want to know what kind of object an elephant is. Each touches a different part of the elephant. The one who touches a leg says 'It is a pillar', the one who touches the trunk says 'It is a pipe', and then the one who touches an ear says 'This is a winnowing fan', and so on. Thus, each opinion differs. Hence, if we wish to understand what kind of object the elephant *is*, we must look at it from all sides. In the context of this elephant illustration, *Pramana* ('comprehensive right knowledge', or complete knowledge) is involved in the 'direct' section of sensory observation, namely, the use of touch. Each blind man forms an example under the category of Naya.

This example conveys an impression of the Jain Holistic Principle. We can now apply this principle to a real larger question. Consider the following seven Naya:

1. Earth may be round.
2. Earth may not be round.
3. Earth may, or may not, be round.
4. Earth may be of indeterminate shape.
5. Earth may be round or may be of indeterminate shape.
6. Earth may not be round or may be of indeterminate shape.
7. Earth may or may not be round, or may be of indeterminate shape.

From these conditional predicates, we reach the conclusion that the Earth is round from a global standpoint but is not round from a local standpoint. A similar conclusion may be reached about Mars and Venus. Therefore, the same may be true for all the planets.

Applying the syllogism to a new planet which has these same properties, we may conclude that this planet is round from a global standpoint but is not round from a local standpoint. Both are 'true', but they are also aspects of a larger truth. We can also speak of a 'local' truth that is a subset of a global or *universal* truth. The 'local' observer who does not experience the roundness of the Earth cannot invalidate the observed truths of astronomers, astronauts, geologists or for that matter anyone who travels any distance! Yet the experience of the local observer cannot be cancelled out by the larger, scientific truth and even experts for the most part ignore the roundness of the Earth when they are not observing or directly experiencing it.

Thus we arrive at the Conditional (non-absolute) Holistic Principle of Anekantavada. The conditional predicates applied to each entity are beads which are held together by the Holistic Principle behaving like a thread.

## The Relevance of *Anekantavada*

The Jain Holistic Principle allows for a more rounded and less exclusive view of the truth than that which has tended to prevail in most organized systems of thought. It acknowledges that several apparently contrasting or even contradictory propositions might be simultaneously 'true', or represent valid aspects of the truth. Equally, it works on the assumption that there are many ways of observing the truth or working towards it. We are all on a journey towards the same destination, but we follow different pathways and are at different stages on our journey. Anekant does not posit that 'all truths are equal' or that all facets of the truth are of equal size and significance. However, it does remind us that we should consider the local and the particular as well as the universal, the intuitive sense as well as the dispassionately recorded fact. It asks us to remember how much we *do not* know and so to respect the knowledge and experiences of others. The foundation on which Anekant rests is the idea of the *Jiva*, or unit of life, as unique and special, each one possessing the potential for omniscience. Each embodied life form and each species has its own 'viewpoint' that deserves to be understood.

Anekant therefore has radical implications for the way we think, behave and live together. The latter process – 'living together' – encompasses the local or national and the global human communities. More than that, it includes the way in which we as human beings accommodate ourselves with the 'rest of nature': the species and ecosystems of the Earth, to which we are intimately linked. The Jain Holistic Principle is a thought process that involves questioning every assumption, every certainty, considering not only *what we believe* (individually and collectively) but also *why we believe it*. The intention behind the belief or ideology is therefore as important as the tenets of the belief system, however lofty these might appear to be. What we call Jainism itself is by no means immune from Anekant. It is itself a belief system practiced by flawed human beings (i.e. men and women who have not reached *Kevalajnana* or omniscience). At the same time, Anekant is an aspect of the Jain sensibility that could be translated to other cultural contexts and in the process strengthens and reinforces them. Anekant is radical in the literal sense of examining every problem from the roots upwards rather than imposing solutions from above.[1]

Therefore it follows that the principle of Many-Sidedness provides an inoculation against extremism or *Ekant*, the one-sided world view that refuses to conceive of other possibilities or approaches. Ekant can occur at all points on the political spectrum and crosses the boundaries of faith. It also manifests itself in mechanistic or reductive approaches to scientific research. In essence, Ekant is the shutting down of our critical faculty. When this happens, we cease to ask questions of ourselves and become unable to understand and sympathize with others. This process leads to the perverted or distorted world views that are at once products of heavy karmic material and generators of new karmic inflow.

The result of Ekant is at best merely an intolerant cast of mind, at worst intellectual and physical violence or Himsa. There is a connecting thread between one-sided thinking and all types of violent behavior. Anekant is a training of the mind which serves to avoid ideological fixations and to keep the channels of communication open to alternative perspectives. It also restores the sense of wonder which is viewed as essential to the pursuit of knowledge. The Jain Holistic Principle can also be called 'Non-violence of the mind'. Contrary to some popular assumptions, it offers a *more* rigorous and inclusive intellectual process than adopting positions of rigid certainty (see Rankin, 2010).

Many-Sidedness asks us to reach beyond political ideologies of 'left' and 'right' and to look for the universal spiritual essence behind the superficial differences between faiths. At the same time, it helps people with different beliefs and cultural backgrounds to find common ground and interact peacefully. In this sense it is the first 'interfaith' tradition. It has enabled Jain communities to accommodate themselves with larger faith traditions (and sometimes learn from them) rather than being absorbed or suppressed.

The Holistic Principle of Jain logic dovetails neatly with current trends in scientific and philosophical speculation that have been gathering momentum for the past half century. In *The Logic of Scientific Discovery* (1968), for example, Karl Popper expresses the view that there can be no 'absolutely true' scientific laws because our understanding is constantly developing and shifting, sometimes radically, sometimes at more subtle levels. This humanistic perspective is, in effect, a western counterpart to

Anekantavada. This principle of uncertainty and non-absolutism has subsequently become the only (almost) fixed assumption of quantum physics! However, the most appropriate last word here should go to the great Indian statistician Prasanta Chandra Mahalanobis (1893-1972). He is best known for the Mahalanobis Distance, a multivariate distance measurement that takes into account underlying patterns and correlations, in other words a many-sided approach to measurement. Mahalanobis founded the Indian Statistical Institute and in a paper on statistics in 1954 wrote of Jainism in the following terms:

> *I should (also) draw attention to the realist and pluralist views of Jain philosophy and the continuing emphasis on the multiform and infinitely diversified aspects of reality which amounts to the acceptance of an 'open' view of the universe with scope for unending change and discovery.*[2]

## Chapter 3

# Truth 1: The Jiva and Karmic Matter

Axiom: *'The soul exists in contamination with karmic matter and it longs to be purified.'*

*Know that the world is uncreated, as time itself is. ... Uncreated and indestructible, it endures under the compulsion of its own nature.*
Jinasena, *Mahapurana*[1]

### Jiva and Ajiva

In Jain cosmology, the 'inhabited universe' or *Lokakasha* is composed of two distinct parts:

1. Non-living material
2. Remainder, i.e. 'living part'

The living part of the universe is described as a *remainder* because it is the vital element, the key ingredient that is also the rarest and least hard to discern. More significantly, it is described as 'pure soul', whereas the non-living material (or 'non-pure part') is described as *karmic matter*. An appropriate analogy is gold ore: karmic matter corresponds with the dross and the left-over twenty-four carat gold may be likened to the 'pure soul'.

However, it is important to be aware that in the Jain universe the pure soul is a non-material entity, but it is encased in matter with which the karmic process has brought it into contact. Nor is 'soul' a simple collective entity or inert mass. Instead, it is composed of individual souls, each of which is equally alive and in search of eventual liberation from karmic entanglement. Thus the individual's search for spiritual liberation and true self-knowledge is the same as the soul's physical journey towards liberation from karmic matter.

In this context, karmic matter is actual physical material that makes the soul 'impure', in other words encases it and cuts it off from a full understanding of itself. The term *karma* has become broadly familiar in the West through increased understanding of Hindu and Buddhist teachings. There, karma is a universal law of cause and effect: 'what goes round comes round'. The word itself denotes action, with each action generating its own reaction. All the workings of the universe, including the cycle of birth, death and rebirth and the destiny of the individual are determined by the law of karma, which is viewed as abstract and impractical.

The Jain perspective differs profoundly here from the other two Indic traditions. Jains do not perceive karma as an abstract law but a physical process which, far from being intangible, touches everything in the universe and directly affects each individual life form. Jains rarely use the term 'karma' as a noun on its own, except in highly specialized treatises such as the *Karma-grantha,* of Acharya Devendra Suri, the Svetambara monk and teacher of the twelfth century CE: 'Acharya' means spiritual master. Instead, Jains speak far more frequently of karmic types, for example *mohaniya-karma,* meaning the 'bliss-defiling' or 'deluding' variety of karmic influence. Nonetheless, they are far more likely in practice merely to say 'mohaniya', with the karmic aspect understood. Jains emphasize the term 'karmic', as in *Asrava,* the influx of karmic particles to the soul, and this can result in loss of self-awareness. They will talk in general terms of karmic matter and the ways to reduce or overcome its disruptive effects. Alternatively, they will speak in diagnostic tones of the *form* of karmic matter (or karmic particles) that is influencing an individual's life, and hence how he or she might reduce its impact.

Jains always draw careful distinctions between different types of karmic matter and the gravity of their effects. Some types of karmic matter present less serious impediments to knowledge than others, and there are also types of karmic influence that, although limiting in some ways, have benign effects and point in positive spiritual directions. Therefore, when Jains speak of karmic influences, they always show an awareness that there are many types of karmic matter and many means and processes by which it influences the soul – and hence the life of the individual associated with that soul.

What all forms of karmic matter have in common is that they are inherently *limiting*. They encase the soul and impose on it physical limits, including limitations of life span and hence entrapment in the birth-death-rebirth cycle (*samsara*) until liberation takes place. In its many forms, karmic matter restricts understanding and obscures knowledge. This is why the Jain spiritual journey is identified with a struggle to overcome the effects of karmic matter and eventually to shed the particles of karma in which the pure soul is encased. By this means, spiritual 'victory' is achieved, the true self is realized and ultimate knowledge attained.

Translated into 'everyday' human terms, this means reducing as far as possible the material attachments that lead to false understanding and warped priorities. It means reducing excessive consumption of resources, having respect for the environment (and the living beings of which it is made), acting with compassion and respect towards all life, i.e. everything that contains pure soul. Of course, all actions on the material plane are karmic. They are products of karmic matter and generate more of it. The pure soul, freed from and fully independent of karmic matter, no longer needs to act. Reducing unnecessary actions and following the principle of Careful Action by avoiding harm to others will reduce karmic influence and eventually enable the shedding of karmic matter. 'Action', in this context, encompasses thoughts and ideas and the mental attitudes they engender. However, the experience of recognizing karmic types

and responding to their effects is always expressed in physical rather than abstract terms.

This is far more than a semantic difference between Jainism and other Indic philosophies. It demonstrates that Jains envisage karmic matter as a substance that occurs throughout the inhabited universe. It exists in the form of aggregates of particles which, in modern scientific parlance, can be described as sub-atomic. Karmic particles interact with the soul and this encounter creates all the complexities, and limitations, of material existence. Like sub-atomic particles, they are not immediately perceptible to us, because the workings of karmic matter *in themselves* obscure our knowledge and vision. Yet we are able to study carefully their effects and speculate on their workings. The Jain viewpoint also posits a direct relationship between the individual and karmic matter, whereby he or she can learn to limit its influence or 'rise above' it.

The encounter between the soul and karmic matter is a natural process. Or, more precisely, it is a series of natural processes by which different types of karmic particles are attracted and exert varying degrees of influence. That said, entanglement with karmic matter is not viewed as the most desirable 'natural' condition for the soul and nor is it the desired 'outcome' of its existence. Much as the most desirable natural outcome for a sapling is to mature into a tree, or a child to grow into an adult, the soul is *intended* to work its way through the many stages of karmic influences to achieve its fully evolved state, free of karmic matter.

Jain teachings take a long view of the soul's journey. It can take place over many lifetimes, with each lifespan itself determined through the workings of karmic matter. The soul's journey is likely to cross boundaries of gender, plant or animal species (and even mineral type) as well as ethnic origin and social class. The spiritual development of the soul may thereby encompass the entire experience of biological evolution, as well as the genetic inheritance of each of the soul's embodiments. Moreover the process of freeing oneself from karmic influence can last for aeons. Realizing this is, in itself, spiritually beneficial, as it offers a useful perspective on the attachments and cravings of mundane existence, making them seem trivial in the overall scheme of things. The experience of meditating on the nature of karmic matter – and beyond that the true nature of the soul – is akin to the cosmologist's understanding of the universe in its vastness, including awareness that there is still much that has yet to be understood.

Another aspect of the Jain view of the soul and its karmic embodiments is an inclusive view of the self. From the western standpoint, both in theology and secular political thought, this can seem especially unfamiliar. The conventional western view of the self is as a non-transferable unit of life. Therefore, the idea of the 'individual' is restricted to one being at a time. The nature of that being is influenced by factors including genetics, evolution and environment, but it is nonetheless self-contained, its experience and consciousness quite literally dying with it. Jain doctrines present a radical challenge to this viewpoint, to an even greater extent than other Indic traditions. This is because of the nature and centrality of karmic matter in Jainism, and because of the Jain view of the soul as a unique life monad. In this there are clearly discernible parallels with the western perception of the individual as unique and (in the words of the American Declaration of Independence) 'endowed with unalterable and inalienable rights'. Jains strongly uphold this view and yet in a crucial respect they enlarge upon it.

According to the Jain perspective, true 'life' is not found in the material embodiment that 'lives' and 'dies' according to karmic influences, but in the underlying immaterial soul that is journeying towards liberation. It is the soul that is meaningfully alive. It is also identical with the true self. Thus, although the soul is a unique and individual being, it has incarnated in a myriad physical forms and species types, all of them generated by its karmic involvement. The experience of each of these embodiments carries forward to the next incarnation, indeed often determines it, and has a long-term bearing on spiritual development.

From this it follows that the awakening self (i.e. the human embodiment aware of the possibility of *Moksha* or liberation) realizes that the soul's experience can include all life forms at all stages of evolution. This leads in turn to a sense that all forms of life (that is, all embodied life forms on the material plane) are connected intimately and reflect aspects of one's inner self. From there, we can arrive at a scientific understanding of the intricate connections between life forms within the material world and the rest of the inhabited universe. Each one is an embodied soul, encased in karmic matter. Each one might represent an aspect of our own soul's past or possible future.

The substance that connects all life forms, all ecosystems on Earth and in the wider universe is karmic matter. It is the point of common origin for all embodied life and in that sense can be called 'origin of species'. Through the interplay of the soul and karmic matter, the connections between all embodied life forms begin and are then perpetuated. For Jains, this insight is at once spiritual and ecological. Self-realization is achieved by respecting the integrity of all other 'selves' in the known universe and recognizing their shared experiences. Such a change of consciousness results in a change of attitude and behavior, in particular a resolve to reduce harm to other beings and minimize exploitation of the environment. This includes forming non-exploitative relationships with fellow humans based on co-operation, compassion and trust, and realizing that human intelligence confers responsibilities to the rest of nature.

Furthermore, Jain practice requires an acceptance that all fellow men and women, indeed all species, are involved in the same spiritual journey. Their beliefs, experiences and perceptions therefore need to be respected, even when they are being questioned or challenged, because only the fully liberated souls understand the whole truth of the universe. In this way, a method of scientific and spiritual inquiry gives rise to an ethos of tolerance, compassion and treading lightly upon the Earth. The Jain sensibility is defined by this system of values, which also point to the behaviors and modes of thought most likely to reduce karmic influences.

In the Jain universe, it is the soul, which is immaterial, that is truly alive, and so it functions as the animating principle of the matter in which it is embodied. Even karmic particles themselves contain souls. The living part of the universe is referred to as Jiva, that which is alive, and the material part as *Ajiva,* that which is not alive. By extension, the word *Jiva* is applied to each individual soul. 'Life' as we understand it within the Lokakasha is based on the interplay between Jiva: the soul, which is genuinely alive, gives temporary life to matter.

There is a sense in which the term 'soul' is a misnomer in the Jain context. Its usual connotation, in most of the world's great faith traditions, is an entity that in some way 'survives death' rather than being the first principle of life. However, we use the term here interchangeably with Jiva, because it is also traditionally associated with eternity. In Jain doctrine, Moksha places the liberated soul beyond the contours of time. Transcending the samsaric cycle (birth-death-rebirth) confers eternal existence in the same way as liberation from karmic matter confers full knowledge of the self and the cosmos. The word 'soul' is also used ambiguously in any description of Jain cosmology, because it can mean either 'pure soul' (unaffected or corrupted by matter) or 'contaminated soul' (weighed down by karmic influences). When we use the term below, we hope that its meaning will readily emerge from the context.

The spiritually aware Jain longs for freedom from worldly

attachments and strives to live in ways that gradually reduce them. In parallel to this, the Jiva longs for liberation from karmic entanglement. This longing is 'hard-wired' into the soul, much as spiritual awareness is an inherent part of the human condition. Yet also hard-wired into the soul is the energy that causes it to vibrate. These vibrations or 'activities', known as *yoga* to the Jains, are the force that first brings the Jiva into contact with karmic matter, and so the cycle of embodiment begins. Moksha occurs when the soul ceases to vibrate. That is why the abolition of karmic influence is equated with the absence of movement. The gradual diminution of such influence is equated with careful, measured actions and *Ahimsa,* the avoidance of harm to others, the guidelines for daily living.

When the soul is liberated and ceases to vibrate, it rises to a higher cosmic level, where all other liberated Jivas reside: the implications of this phenomenon will be considered below. As one soul achieves liberation, another comes into existence, begins to vibrate and encounters karmic matter. Both the emergence of the new soul and its subsequent 'contamination' take place spontaneously. Jain teachings acknowledge no 'First Cause' or divine creator. Instead, they express a primal awareness that 'energy can neither be created nor destroyed', from which a complex cosmological scholarship has emerged over many centuries. Jain cosmology views the universe as eternal and cyclical. It is divided into *Utsarpini* and *Avasarpini,* respectively the upward and downward movements of the eternal cycle of time. The divisions between these cycles are likened to the spokes of a wheel. Each cycle can last for thousands of millions of years as the universe waxes, wanes and regenerates itself – and the liberating doctrine of the Jinas is continuously lost and recovered.

These ancient insights are finding common cause with modern cosmology, which is coming to emphasize increasingly the importance of cycles, as opposed to the once dominant linear patterns beloved of the European Enlightenment. The move from linear to cyclical thinking is also a strong component of modern physics and the natural sciences, finding echoes in philosophy and the visual arts. This is another respect in which the scientific spiritual system of the Jains resonates powerfully with the twenty-first century *Zeitgeist*. Jain logic and methods of reasoning can make a powerful contribution to a wider process of questioning of received wisdom in science and other areas of intellectual life, including politics and sociology. The Jain theory of interaction between Jiva and karmic matter is a spiritual counterpart to many emerging theories of how the universe works. It also has radical implications for the way we conduct our lives.

## Jain Theoretical Science

To understand the interaction of the Jiva or soul with karmic matter in Jain doctrines, it is first necessary to 'think like a Jain'. This means working through the various formulations used by Jains over millennia to understand how the universe operates. By doing this, we may draw conclusions about the compatibility with, and radical differences from, the modern scientific world view.

We are aided greatly in this endeavor by the fact that Jain teachings combine the scientific and spiritual world views. They draw no distinctions between these two methods of looking at the cosmos, but see them as part of a continuum, with 'spiritual' and 'scientific' truths ultimately identical. The investigations of the scientist and the spiritual seeker into the nature and meaning of reality are viewed as two sides of the same coin or two pathways to the same summit. Jain teachings, moreover, are transmitted by a powerful corpus of scholarly research based on reasoned thought, which are subject to continuous and rigorous criticism, examination and addition. There is a parallel tradition that is largely oral and intuitive, but once again these two forms of understanding reinforce each other.

Jain doctrines themselves and the wider culture we have identified as *Jainness* both place a strong emphasis on the pursuit of knowledge through reason. To understand the Jain thought process, we shall speak in terms of two scientific approaches: *Jain Theoretical Science* and *Jain Applied Science.* Jain Theoretical Science corresponds to the first three (spiritual) Noble Truths and (scientific) Axioms. These explain the workings of karmic matter and its interaction with the soul from a theoretical perspective combining physics and philosophy.

This approach leads us logically to Jain Applied Science, which corresponds to the fourth Noble Truth/Axiom. Divided into three parts, this Truth explores the ways in which the spiritual seeker (i.e. the spiritually aware *and* rational human being) can attract or repel karmic influences, understand the workings of karmic matter and work towards freedom from it. The perspectives used will correspond at least as much to the social sciences and psychology as the natural or physical sciences.

In the sections below, we use a combination of 'scientific' terminology (language familiar to modern scientific researchers), 'Jain' concepts (concepts familiar to practicing Jains) as well as language that unites the two approaches.

## The Nature of the Jiva

The idea of the soul or Jiva can be summarized as follows:

In nature there exists a non-material substance composed of four main properties[2]:

Knowledge – *Jnana*
Perception (or *conation:* voluntary action; the desire to perform an action) – *Darshana*
Bliss – *Sukha*
Energy – *Virya*

The non-material substance is the soul and it can be divided into a myriad of individual 'souls' generated by the energy inherent in the universe: the cosmic creative process.

We shall call the four main properties of the soul the *soul elements*. In traditional Jain theory, they are known as *Guna*. The first two of these elements are cognitive functions of the soul and represent 'consciousness'. Bliss (Sukha) is a state that includes 'compassion' and 'total self-sufficiency'. This means the state of compassion for all beings that arises from knowledge and self-awareness. Self-sufficiency in this context means a sense of equilibrium and full consciousness of the self and the universe. It is the opposite of the state of karmic bondage (see below) with the physical and emotional dependence that stems from it.

Energy or *Virya* is an abstract force which powers the operation of the knowing and perceiving qualities of the soul. Unlike matter, it can neither be created nor destroyed. It is the power underlying the creation of 'new' souls, the encounter between the soul and karmic matter and the soul's journey towards Moksha or liberation.

As we have seen, one of the principal words used for soul in Jainism is Jiva, which means 'living part'. The soul therefore represents the aspect of the universe that is truly alive and conscious.

## The Concept of Karmons

Karmic matter consists of sub-atomic karmic particles, which we shall refer to here as *karmons*. These karmons float freely and randomly in space but they do not interact with each other. We may presume that the gravitational force is very small, although that question is not directly addressed by traditional Jain theory. Among sub-atomic particles, karmons are unique in the sense that they can only be absorbed by the soul, and cannot fuse by themselves. In other words, karmic matter as 'molecules/ aggregates' of karmons can exist, but only when they are in conjunction with the soul. Thus karmic matter increases by absorbing new karmons and decreases by dropping some of its karmons in space.

## Interaction of Karmic Matter and Jiva

The Jiva or soul in its purest state has infinite knowledge, perception, bliss and energy. The soul is sentient energy but in general, as the Axiom states, the embodied soul is *polluted* by the karmic matter. The interaction of two highly contradictory elements, soul and karmic matter, can lead to severe distortions of the principal soul elements.

In particular, the presence of karmic matter:

1. *Obscures* the knowledge element of the soul;
2. *Obscures* the perception element of the soul;
3. *Defiles* the bliss element of the soul;
4. *Obstructs* the energy element of the soul.

These interactions between the soul and karmic matter in turn affect the actions, ability and character of the individual, of whatever species or genus. The individual, after all, is an *embodiment* soul, i.e. the product of the soul's encasement in a material form through its contact with karmic matter. The influence of karmic matter prevents the soul from realizing its pure qualities. However, a human being is intellectually and spiritually evolved enough to achieve full understanding. This takes place through reason, meditation and conscious changes in behavior, attitude and priorities.

Sukha, or bliss, undergoes the most profound alterations as a result of karmic influences. Whereas the other three soul elements are merely obscured or obstructed, Sukha is *defiled* or *polluted*, meaning that its nature is changed and that it needs to be purified or refined by the shedding of karmic particles. A helpful analogy might be the changes that take place under the influence of intoxication by alcohol. This leads to a wide range of changes in behavior and mood, with normality restored only when the physical and psychological effects of alcohol wear off. Levels of alcoholic influence can be measured through the blood, alcohol and breath: its presence temporarily alters the equilibrium of the body. The same is true of *Mohaniya*, the 'bliss-defiling' karmic component.

This process of *defilement* or quasi-chemical change in the Sukha or bliss element of the soul goes on to affect the Virya or energy element, by *perverting* it towards false perceptions and ignorance in place of knowledge.

Karmic matter can only survive when it is involved or interacting with the Jiva. However, the Jiva is self-sustaining and has an inherent tendency to seek freedom from karmic matter, including embodiment. This inherent tendency of the soul will be referred to here as the *freedom-longing catalyst*.

In traditional Jain science, the freedom-longing catalyst is known as *Bhavyatva*, which is described as a sense of the possibility lying dormant within the soul until the moment when it is triggered. The trigger effect is produced by that soul's particular interaction with karmic matter, in this case 'light' or 'positive' material. At the same time, it is the catalyst for liberation from karmic bondage.

## The Language of Karmic Influence

We now describe the *ways* in which karmic matter interacts with the soul, or how Jiva meets Ajiva. To do this, we use a series of technical terms, each of which forms a sub-section below. As with the previous section, we have used a fusion of modern scientific and traditional Jain terms to illustrate the continuities between the two modes of thought. The terminology or *language of karmic influence* used here is intended to increase understanding of Jain Theoretical Science – by 'translating' spiritual into scientific terms and vice versa.

## The Karmic Process

The bond that forms between Jiva and karmic matter is called *karmic bondage*, or *Bandha*. For convenience, we often speak of 'contact with karma', but in Jain theory that is not envisioned as direct or tactile contact, but as the positioning of karmons around or alongside the soul, exerting influences both overt and subtle but not fusing with the soul itself. P.S. Jaini describes the relationship in these terms:

> *It should be made clear that Jainas view the souls' involvement with karma as merely an 'association'* (ekakshetravagaha, *literally, occupying the same locus); there is said to be no actual* contact *between them.*[3]

Although the karmic matter does not directly touch the soul – because the soul is a non-material substance – the presence of karmons around the soul exerts influence and can alter the soul's character. Karmic matter, coupled with the soul's perverted energy element, produces a *karmic force-field* (or *karmic field*). In turn, the force-field gives rise to *karmic influx*. Both these effects are part of an aspect of reality known to the Jains as *Asrava*. This is the flow of karmons from all directions into the soul. Further, the karmic force coupled with the soul's obstructed energy element fuse the incoming karmons: we will call this process *karmic fusion*, but in traditional Jain terms it is an aspect of *Bandha*. The total karmic matter fused to the soul is thereby revised, and this dynamic karmic process continues.

It is important to remember that there are two distinct but interconnected aspects of Bandha: *karmic bondage*, or the actual physical condition that allows for the bombardment of the soul by karmons, and *karmic fusion*, or the actual assimilation of karmons with karmic matter as a whole. That means that the particles cease to be isolated units but become part of something larger, namely a type of karmic matter that exerts influence on the soul. Where there is no larger unit of karmic matter, karmons can exert no influence of either negative or positive kind.

As well as karmic fusion, there is a parallel process of *karmic decay / karmic fission*, or the shedding of karmic material. This takes place when karmons drop out or are emitted, as a result of which the karmic matter loses its density and its power – and the type of influence it exerts can markedly change.

## Karmic Density

Karmons exist in an undifferentiated form in nature, but the karmic force-field coupled with the obstructed soul energy introduces specific functions to the karmons, so that they are differentiated. The result is a variety of 'karmic types', exerting a wide range of influences over the soul. As Jaini explains:

> *Karmic matter is said to be found 'floating free' in every part of occupied space. At this stage it is undifferentiated; various types (prakruti) of karma, classifiable by function, are molded from these simpler forms only after interaction with a given soul has begun.*[4]

It is assumed that the karmons recompose into 'heavy or light karmic matter', i.e. karmic matter with high or low density. *Heavy karmic matter* implies that the karmic bondage is strong, whereas *light karmic matter* implies that the karmic bondage is weak. It is therefore easier to remove light karmic matter from the soul. Thus there is a dynamic process of updating the karmic matter and hence its functions.

The karmic density distinguished above as light or heavy karmic matter depends on the following factors, expressed below in secular scientific and Jain spiritual terms:

1. The number of karmons involved in karmic fusion –

*Pradesha;*

2. Different types of karmic matter / karmic components – *Prakruti;*
3. The potential energy in karmic decay – *Anubhava;*
4. The time that fused karmons take to decay – *Stithi.*

These four main karmic components are the antithesis of the four principal soul elements. Therefore, they defile the 'bliss' element, obscure the 'knowledge' and 'perception' elements and obstruct the 'energy' element of the Jiva. The factors 1-4 listed above also represent the order in which karmic fusion takes place.

'Heavy' and 'light' karmic matter is also frequently referred to as 'negative' and 'positive' karmic influences. The interchangeability of these terms within Jain culture reflects the sense of continuity between scientific and ethical approaches. This is because the Jain Dharma is a natural or universal law that encompasses both ethical and physical processes, seeing them as ultimately one and the same. For the soul, the correct workings of Dharma point towards eventual liberation through the process of karmic decay. The terms 'heavy' and 'light' refer to shade as much as to weight. That is to say, they describe the extent to which karmic influences obscure the Jiva's knowledge and self-awareness by darkening its vision. Thus the *lightening* (and eventual disappearance) of karmic matter is identified with increasing *enlightenment*.

In Jainism, the actions and mental attitude of individuals have a powerful bearing on their karmic destiny. The conscious choices we make can transform heavy karmic matter into light karmic matter, and thereby radically alter the effects of karmic involvement on the soul. The lightening of karmic influence gives rise to further enlightenment and eventually to Moksha as the obstacles to understanding are peeled away.

## Long-term Equilibrium State of the Jiva

We have described above the short-term state of the Jiva or soul under karmic influence. Now let us turn to the soul's *long-term equilibrium state*. When all the karmic matter is removed from the soul through the emission of karmons, what is left is the pure soul, that is, it has infinite levels of the four elements of the soul described above.

There are two stages in the attainment of this state of consciousness, which is the ideal 'natural' state most in keeping with the Dharma: the zenith of spiritual evolution. In the first stage, karmic influx is blocked through the creation of a *karmic force shield,* which implies the total end of the inflow of new karmons. The next stage is the shedding of all accumulated karmic matter, which gathers momentum and eventually results in *total karmic decay/fission.* Jains call these two processes of stoppage and shedding *Samvara* and *Nirjara* respectively.

When all the karmons have been emitted, the soul no longer has a karmic field and thus has attained its full potential. This is the *liberated state* in which the soul has transcended karmic matter and so neither generates nor repels it. Until this happens, *Asrava* and *Bandha, Samvara* and *Nirjara* take place continuously as cycles of fusion and bondage, stoppage and fission.

In Jain Theoretical Science, the idea of karmons has a profound impact on the workings of the physical universe, the evolution of life and human psychology. The latter is but a small subset of the first two phenomena. However, it is considered to be a crucial component: this is because of the extent of human creativity (and destructiveness) coupled with the human capacity for spiritual insight (and delusion).

The idea of karmons also demonstrates that Jain theory has, over more than two millennia, developed a sophisticated understanding of sub-atomic particles and their effects on the way the universe works. The insights of the Jains in this area of science stem from a combination of intuition, reason and scholarly speculation. We may call this approach *spiritual-rational* to differentiate it from the *secular-rational* approach of the orthodox scientific model that has been dominant for the past 500 years. Scientists working within that secular-rational framework have only in relatively recent times begun to come to terms with these particles and their significance. By so doing, orthodox science has started to challenge its own tradition of linear thinking and become more aware of cyclical patterns in nature and the complex interplay between living organisms. This is, in effect, a convergence with the idea of Dharma as understood within Jainism.

## The *Tattva* or 'Nine Reals'

Jain cosmology is holistic, in that it emphasizes the connections between all forms of life in the universe: this 'interdependence' functions on both ecological and ethical levels. However, it places equal emphasis on the unique identity and qualities of each individual life form containing 'soul' or Jiva as its animating principle. From this it follows that even the most apparently elementary forms of life (from sub-atomic particles to so-called primitive organisms) can play complex and critical roles in nature. They may also be crucial to the continuing health and survival of supposedly higher species, including humanity.

These conclusions are being embraced increasingly – albeit cautiously – by secular science. In Jain teachings, they operate at the 'scientific' level, governing the material element of the universe, and the 'spiritual' level, governing the non-material element of the universe, in other words the Jiva or soul. The basis for Jain cosmology is an understanding of reality based on nine elements known as the *Tattva* or 'Reals' that reflect the interplay of the Jiva and karmic matter. Many of the concepts referred to in this section have been mentioned above, but through the Nine Reals they become parts of a coherent whole. The *Tattva* are also fundamental to the way in which Jains think and what they conceive of as reality.

For the purposes of Jain science, the Nine Reals can be rendered in two ways, the first corresponding primarily to the physical laws of karmic influence, and the second to the traditional spiritual insights familiar to practicing Jains. The physical order is rendered as follows:

1. Karmic matter
2. Karmic bondage/fusion
3. Karmic force/influx
4. Karmic force shield
5. Karmic decay/fission
6. Liberation
7. Jiva/Soul, surrounded by:
8. 'Heavy' karmic matter
9. 'Light' karmic matter

This order expresses the Jain view of reality in terms that can be recognized by the scientific researcher in the laboratory, working through physical or chemical processes and arriving at conclusions about them. The first five categories refer to the material (Ajiva) aspect of the universe, the last three to the non-material soul (Jiva) in conjunction with Ajiva. The sixth, Liberation (Moksha) is the point at which the Jiva breaks free from Ajiva. Thus the non-material substance, freed from karmic weight, rises to a higher level of the universe reserved for pure consciousness.

Yet at the same time that this happens, a new soul is embodied, i.e. new Jiva substance comes into contact with karmic matter. Category 7 meets categories 8 and 9 and the process of categories 1-6 begins anew. For the scientific researcher, karmic matter would be the most likely starting point, for its dynamics govern the organization of the 'inhabited universe' (Lokakasha), the area of the cosmos where embodied souls reside. The Lokakasha is composed of six 'existents' (*Dravya*), the role of which is explained below (see Chapter 5).

The order of the Nine Reals traditionally understood by Jains takes the Jiva as its starting point. From the perspective of spiritual development, the non-material, 'pure soul' aspect is the most important, as both the origin and fulfillment of life. Rousseau famously declared of the human condition that: 'Man is born free and is everywhere in chains.' The same is more or less true of the Jiva. It arises out of pure energy, which makes it vibrate and thus embrace chains of karmons. Its subsequent spiritual quest takes it towards liberation from karmic bondage, when it becomes a *Siddha* (liberated soul) that has attained the pure consciousness natural to it. Therefore Jain practice usually renders the *Tattva* in this order:

1. Jiva = Soul, non-material aspect of universe
2. Ajiva = Material and insentient aspect of universe including karmic matter
3. Asrava = Karmic force/influx
4. Bandha = Karmic bondage/fusion
5. Punya = Light/positive/auspicious karmic matter
6. Papa = Heavy/negative/inauspicious karmic matter
7. Samvara = Karmic force shield / stoppage of karmic inflow
8. Nirjara = Karmic fission/decay/shedding of karmons
9. Moksha = Liberation/self-realization/release from karmic influence

The first order of the Reals corresponds most closely with physical or biological evolution, the second with spiritual evolution, but the two evolutionary forms are intertwined.

Jains regard the Tattva as existing eternally (i.e. neither created nor destroyed) and use them to explain the evolution of the universe. As Archarya Jinasena wrote in the ninth century CE:

*Know that the world is uncreated as time itself is,*
*Without beginning and end,*
*And is based on the principles, life and the rest* (See Endnote 1)

The 'world' here can be taken to include the universe. The 'principles' mean the Nine Reals, 'life' means the soul and 'the rest' implies the other eight Reals (i.e. the eight aspects of Ajiva). The Reals of Jain science take the place of any single First Cause or creator god. They include all the creative forces within the universe, including those that elude full human understanding. Thus meditating upon or seeking greater knowledge of the Nine Reals is the purpose of Jain spiritual practice, rather than worshipping a Supreme Being or deity. For these reasons, Jainism is sometimes called a trans-atheistic religion, as opposed to either a theistic or atheistic belief system. Jains can also acknowledge the existence of deities but these do not represent motive forces within the universe or even necessarily higher levels of consciousness.

One of the reasons advanced by Jain teachings against the idea of a supreme deity or creator god is that such a being would be responsible for creating imperfection and suffering. Therefore, in terms of spiritual evolution, such a being would not be a perfect soul! Furthermore, a higher being would only *wish* to create a perfectly ordered universe. Jains also deploy an argument that has become classic among rationalists and humanists since the European Enlightenment, the question of 'Who created the creator?' As the German philosopher Friedrich Nietzsche asked: 'Is man only God's mistake or is God only man's mistake?' The question has been taken up at some length by philosophers and scientists alike. Jinasena, in the *Mahapurana*, expresses the argument in these terms:

*Some foolish men declare that a Creator made the world. …* (But) *if God created the world, where was he before creation? If you say he is transcendent then, and needed no support, where is he now?* (See Endnote 1)

Critics of Jain philosophy often claim that it takes an excessively pessimistic view of the universe. Yet the Jain emphasis on perfection – and perfectibility – denotes a spirit of ultimate optimism. Yes, the inhabited universe is imperfect and life within it is flawed, but the *true* nature of life is balanced, ordered and free. All embodied life forms have the possibility of working towards perfection, even if that means experiencing many embodiments to get there. Unlike other faith traditions, Jainism does not accept the idea of imperfect deities as controlling forces, with characteristics such as jealousy or aggression. Instead, each individual Jiva has the potential to rise to a state of pure reason and knowledge. Jains are therefore not so much pessimists as perfectionists.

In the context of Jain rejection of the idea of a creator god, it is worth noting that Jain teachings speak of an eternal, cyclical universe. Also, the perfect soul or fully realized Jiva possesses omniscience and immortality, compassion and equanimity, attributes that other faiths assign to divine powers. In this sense, there is a 'divine spark' within all life forms: the infinite potential of Jiva.

## Some Scientific Analogies: Magnetism, Petrol, Oiled Cloth, Viral Infection

We have explained karmic theory above through a combination of scientific language, drawn primarily from physics, and spiritual language, drawn from the corpus of Jain doctrines. However, it is also useful to look briefly at some analogies with physical processes familiar to students of science. They are intended to increase direct understanding of the way karmic matter interacts with the soul in Jainism, but they should not be seen as literal parallels or literal reflections of the karmic process.

We may regard the polluted soul as a magnet. It attracts iron filings that can be considered as representing karmons. The magnetic force lines are equivalent to karmic force lines, the joining of the iron filings to a magnet can be looked at as karmic fusion: the filings become strongly bonded to the magnet much as the karmons become strongly attached to the soul. Creating a force shield that stops new filings from being attracted may be seen as a form of insulation. The shedding of old particles in the force-field implies 'demagnetization' so that there is no attraction. When all the particles are dropped, the soul is free from the magnetic element of this karmic matter and what is left is the liberated soul.

A second analogy that springs to mind is petrol. This is a refined stage of crude oil, and only the process of refinement leads to the full combustive power of petrol. Obviously, the refined stage represents the pure soul and impurities are analogous to karmic matter.

The impure soul may also be likened to an oiled cloth. This cloth can attract, because of the moisture, dust particles analogous to karmic particles, with the bond between the cloth and the oil akin to karmic bondage. Note that the nature of the soul remains invariant under adaptation to a particular body's dimensions, like a cloth which can be folded into various shapes without any alteration in its mass.

A final and interesting analogy is with the way a virus can affect its host body, resulting in changes, such as the symptoms of physical illness, in much the same way as karmic matter influences the soul. The viral infection represents karmic matter in this context, whereas the infected organism represents the Jiva or soul.

The philosophy of the Jains can be classified as a form of esoteric science. This is because it is a quest for the knowledge lying within: the knowledge of the soul. It delves beneath those facets of reality that can be observed, monitored and quantified.

Jain Theoretical Science is therefore concerned with subtle, complex interactions between living systems – and beyond that between the material and non-material aspects of the universe. We can view the cosmos from a 'pure' scientific perspective, in which we focus on the interaction of material forces, or from the spiritual perspective, in which we look beyond the material for hidden meanings, connections and patterns. Increasingly, the two viewpoints are converging. They express *aspects* of the truth, rather than the truth as a whole. It is much as light can exhibit the properties of either particles or waves, depending on the way in which it is considered, but it is something more than either of these: it is light!

Through its growing acknowledgement of complexity and interconnectedness, orthodox science is becoming steadily more esoteric in character. Scientists reach out increasingly to ancient spiritual teachings to enlarge their understanding of physics and cosmology. In this environment, the encounter between Jain doctrines and orthodox (*exoteric*) science perhaps holds a special promise. Jain teachings have not been suppressed or interrupted by colonialism or conquest. They have evolved from pre-literate times to suit socially and technologically advanced societies past and present, and they are based on the eternal quest for knowledge.

## *Leshya* or 'Karmic Colors'

The Dharma's emphasis on preserving the environment is emphasized through the 'color coding' of the karmic density of the Jiva. The term *leshya,* literally meaning karmic stain, is used to describe these karmic colorations. There are six levels, forming a hierarchy: black (*krishna*), blue (*neel*), grey (*kapota*), red or yellow (*tejo*), lotus-pink (*padma*) and luminous white (*sukla*). The subtle body to which the karmons adhere fluctuates constantly between these colors according to the type of karmic matter that predominates. At the moment of transmigration, the coloration of the karmic matter indicates the health of the Jiva and so has a powerful bearing on the next incarnation. In a form of spiritual synesthesia, the colors can be tasted and smelled by adepts as well as visualized at the subtle level.

Color as a metaphor for mood or state of being has deep roots in East and West alike. In North America and Europe, color therapy has become an increasingly popular form of holistic treatment for both psychological and physical ailments. Some clairvoyants are able to see the color of a client's aura and draw conclusions about that client's health, well-being and future prospects. In so doing, they are working with an ancient Jain system of karmic discernment. J.L. Jaini (1916), for example, makes an explicit connection between the leshya and the colors of the human aura.

The first three leshya (black, blue and grey) are held to represent heavy karmic density, whereas the last three (red/yellow, lotus-pink, luminous white) represent lighter karmic density. At a popular level, Jains often describe the leshya in terms of the use of a tree and the attitudes that this reflects. A person with the first leshya level (black) uproots the tree for its fruits, the second (blue) cuts the tree from its trunk, the third (grey) cuts off a branch, the fourth (red/yellow) breaks off part of a branch, the fifth (lotus-pink) plucks ripe fruit from the tree and the sixth (white) merely picks up ripe fruit fallen to the ground. This popular interpretation of the leshya is far from fully instructive, but has value in its summary of the different levels of karmic activity. It also highlights the importance of preserving the environment and working with nature rather than seeking to exploit or conquer it. The sixth person in the story preserves every aspect of the tree and does not interfere with it for his or her own ends. Thus the purest (in Jain terms) of karmic colorations is associated with the conservation of nature.

Furthermore, the karmon intake is increased by creating waste and pollution, since these are regarded as acts of violence against both the Earth and the self. Such actions have a strong impact on the karmic coloration of the Jiva, as do obsessive attachments to possessions and worldly trappings. Such attachment is described in the *Tattvartha Sutra* (7.17) as a psychedelic or mind-altering state and a state of trance (*Murcha parigraha*).

In popular Jain culture, another frequent analogy is made between the state of karmic lightness (luminous white leshya) and 'the bee that sucks honey in the blossoms of a tree without hurting the blossom and meanwhile strengthening itself'.

## Chapter 4

# Truth 2: The Jain Hierarchy of Life

Axiom: *'Living beings differ due to the varying density and types of karmic matter.'*

### The Axiom

How does karmic matter divide different living species and forms? According to the second Truth of Jain science, the differing density of karmic matter does much to explain the differences between life forms. In other words, the 'purer' the basic elements of the soul become, the 'higher' the form of life that incarnates. In this context, 'purer' means relatively free of karmic influences, especially of the heavy or negative types. 'Higher' means relatively capable of thinking and acting autonomously, making ethical choices, as well as having the capacity for spiritual insight and the pursuit of knowledge. In this and the following chapter we shall define more fully the *types* of karmic matter, both heavy and light, i.e. the components into which karmic matter is differentiated.

The idea of a *hierarchy of life* would, on first examination, seem to contradict both the popular interpretation of Jainism and the most evident expressions of Jain philosophy. According to the popular view, Jains regard all forms of life as 'equal'. This biological egalitarianism explains the profound emphasis on *Ahimsa:* non-violence and the maximum possible avoidance of harm to all forms of life. The most obvious and therefore most chronicled and photographed form of Jain behavior – the ascetic carefully sweeping the ground before him with his brush – is therefore acting upon the belief that all life forms are equal and so there are no fundamental differences between humans and micro-organisms.

That approach, however attractive or compelling it might seem, is at once a distortion and a reduction of Jain thought. The reality is more subtle and complex. Jains regard all life in the inhabited universe as containing Jiva or soul as its animated principle. Because the soul's natural evolutionary process points towards Moksha or liberation, all embodied life forms have the potential to become liberated souls in possession of full self-knowledge. All liberated souls are equal because they have achieved the same levels of knowledge and are free from the karmic cycle. Therefore, it follows that all embodied life has the *potential* for equality, which can only be achieved through the transcendence of *samsara:* the cycle of incarnations caused by karmic matter.

However, it also follows that within the samsaric cycle there are forms of life at many different stages of biological and spiritual evolution, serving different purposes and equipped with different functions and capacities. There are inherent differences between species in terms of mental capacity and spiritual awareness as well as spiritual properties. Humans have – on Earth at least – achieved the highest capacity for spiritual development, reason and pursuit of knowledge in its purest forms of any species that has so far evolved. They also have greater capacities for independent action and judgment, free will and making choices (rational and irrational) than other species, including deities or *Devas,* whose abilities to act are relatively limited. From this understanding of human strengths, it follows that incarnation in human form is the logical route to Moksha. Becoming human is not in itself a passport to liberation, but a critical staging post on the way to freedom from karmic influence. Thus, humanity is at the pinnacle of the Jain hierarchy of life. *Homo Sapiens* is the most spiritually evolved being and a human incarnation corresponds to relative lightness of karmic matter.

At the same time, Jainism is a philosophy of 'both/and' rather than 'either/or'. The concept of a hierarchy of life is far from a blueprint for human supremacy, as it has often become in other faith traditions and secular philosophies. In fact, it is exactly the reverse. Human and intellectual powers confer responsibilities in place of rights with regard to the rest of embodied life, in other words the natural environment on which humans depend for their survival. Because humans have the ability to act freely and the power to differentiate between benevolent and harmful acts, they are obliged to act carefully at all times instead of giving in to their immediate and superficial desires. Individually and collectively, they are enjoined to question and examine their thoughts and actions in order to minimize harm to all life – their own lives included.

By so doing, human beings are not submitting to an external authority but being true to their inner selves, the Jiva or soul component. They become increasingly conscious of the Jiva as their spiritual understanding develops. When humans act with care and restraint, reducing their ecological footprint and their dependence on material consumption, they are fulfilling their potential far better than when they seek to dominate the rest of nature or for that matter other human groups. This is because they are showing a capacity for critical reflection and moral choice that distinguishes them from other species and brings them closer to Moksha. They are bringing themselves into alignment with Dharma, the law of compassion and interdependence that governs the universe – and which as humans they are equipped to understand.

It is therefore *because* humans have such strong intellectual powers that they have the responsibility to understand and control their harmful impulses. For human intelligence has a corollary of destructive powers more extreme than that of any other life form. The potential for spiritual liberation is balanced by the potential for total destruction, and this is because humans have the ability to choose harm as well as careful action, to embrace ignorance as well as pursue knowledge. Jain ethics and Jain science are constructed respectively to enable humans to live in accordance with the Dharma and to understand the workings of the universe that underlie that way of life. Together, they point towards release from human weakness as much as realization of human strength.

Whereas the materialistic model has tended to emphasize human activity as an end in itself, the Jain model emphasizes human restraint. The Jain model is based on learning to live within limits, instead of continuous material growth. From this standpoint, human restraint is a true indication of intelligence and power. Materialism and the ideologies of growth that stem from it are identifications with Ajiva at the expense of Jiva, the non-living part of the universe instead of the element that is genuinely alive. They are literally 'soulless' and in Jain terms the denial of the soul is really a denial of life. It is also one of the most powerful impediments to knowledge and inducements to the 'heaviest' forms of karmic matter.

Jains are aware of the interdependence of all life at two levels. First, at the material or 'embodied' level, they understand that each living system has a purpose and that the survival of one living system can be critical to the survival of a myriad of others. They understand that even supposedly 'primitive' life, including micro-organisms, can have vital significance. Human intelligence demands of us that we remain aware of such connections at all times and live accordingly.

The second level of awareness is that all life forms are connected by possession of the soul element, the Jiva. Each Jiva is unique and of equal potential value because of its potential for spiritual development. In this way, Jains believe simultaneously in a hierarchy of life *and* a cycle of life. All elements in the hierarchy are intertwined and mutually dependent instead of being perceived as separate and exclusive.

Living systems are also at once equal *and* unequal. They are equal in the sense that they all contain the same animating principle (Jiva). Accordingly, all life forms have the same purpose, the movement towards Moksha. However, life forms are profoundly unequal because their karmic destinies and consequent embodiments (Ajiva) endow them with different capacities and functions. Also, each life form is at a different stage on the journey of spiritual evolution, with some more evolved than others in their capacity for reason and insight. This holistic understanding of nature – hierarchical *and* cyclical, equal *and* unequal – is typical of a Jain thought process that is pluralistic rather than binary in character.

From the Jain conception of a hierarchy of life, we may arrive at two conclusions:

1. The Jain ethos and practice of Ahimsa is based on the intuitive and rational understanding that all life forms (i.e. all stages of the hierarchy) are mutually dependent for their survival. Further, the inequality between embodied life forms is balanced by the equality conferred on them by the possession of Jiva. Therefore, the common ecological threads uniting all living beings are reinforced by a common spiritual bond. Ahimsa reflects awareness of biological and spiritual connections alike, and above that an understanding that human intelligence must be applied with restraint and constant self-questioning.
2. Jain spiritual science is far from antagonistic towards western humanistic science. Instead, it complements the traditional humanistic and empirical approach of the West with an approach that includes the 'spiritual' element in nature – which is equated with the origin of life. This enables it to fill some of the gaps that materialistic rationalism cannot see inside. Western science, after all, is moving rapidly from linear to cyclical patterns of thought and to finding subtly powerful connections between living

systems (or areas of the universe) once believed to bear little or no relationship to each other.

Within Jain thought, there is both the primal sense that 'everything is connected' and the rational emphasis on the value and purpose of each individual life. There is the sense of continuity between humankind and nature, but an equal sense of human intelligence and possibility. It follows that there are many overlaps and intersections between the two scientific approaches, as well as important differences. These commonalities and differences enrich the encounter between the two systems of thought (see Appendix 2).

## Life-units and the Life-axis

The position of each life form in the hierarchy is determined by its relationship to karmic matter. In practice, this means the 'purity' of the Jiva, that is to say its lack of contamination or distortion by karmic particles. The degree of 'soul-purity' can be quantified in a relative way. We may define for convenience a unit of soul-purity as that degree of purity of the soul which leads to 100 life-units in the average human being.

At one extreme, the pure soul will have infinity of units whereas an insentient object will have zero life-units. Thus we can represent the soul's purity or the life-units of the living beings along a line taking value from zero to infinity: we will call this the life-axis. Note that as the degree of soul-purity varies from zero to infinity the density of the karmic matter will vary from infinity to zero, inversely as it were.

## The Role of the Senses

1. The lowest forms of life are the micro-organisms which possess only one sense, that of touch. These are infinitesimal and can only exist as part of a larger body (living or non-living) and therefore they should have very few,

say $10^{-4}$ life-units.

2. The next stage of life is another group of one-sensed micro-organisms which take the subtlest possible unit of matter as their homes and these are earth-bodied, water-bodied, air-bodied and fire-bodied. We will denote water, air or fire-bodied beings by a life-unit of $5 \times 10^{-4}$.

Of these two most basic, yet critically important stages of embodied life, Jaini observes:

> At the very bottom of this scale hence comprising the lowest form of life, are the so-called nigoda. These creatures are sub-microscopic and possess only one sense, of touch. They are so tiny and undifferentiated that they lack even individual bodies; just above the nigoda is another group of single-sense organisms whose members take the very elements – the subtlest possible units of matter as their bodies; hence they are called the earth bodies (*prthvi-aiyika*), water bodies (*apokayika*), fire bodies (*tejo-kayika*) and air bodies (*vayu-kayika*).[1]

This description illustrates the degree of sophistication in the Jain analysis of the origins and evolution of life on Earth.

3. Next are plants, which are rated higher than the preceding life forms, having gross individual bodies, and we will rate them at $10^{-3}$ life-units. It is interesting to note that one can distinguish between the various concentrations of life in plants. For example, onions are believed to have a more concentrated form of life than apples since one seed of an apple gives rise to thousands of apples, i.e. the life gets subdivided, whereas an onion root gives only one onion in the process and, therefore, we can regard the life in an onion as not $10^{-3}$ life-units but something like $10^{-2}$ life-units. This comment also applies to trees. Further, plants or

dead flesh infested by innumerable micro-organisms will also have a higher degree of life-units.

4. When some of the karmic matter is removed then the next highest form of life is that in which the being has two senses, a body and a mouth/tongue. The two senses are touch and taste as it appears say, for example, in seashells, mussels, etc. We give these beings 2 life-units.
5. The next stage of higher life has, of course, three senses; it has also a nose, that is, having the additional sense of smell, for example an insect without eyes. We regard these as having 3 life-units.
6. Further reduction of karmic matter leads to four-sensed beings which develop eyes or the sense of sight, for example, bees, flies, etc. These are assigned 4 life-units.
7. Finally, within the non-human realm, we have beings with ears or a sense of hearing, for example, horses, camels, etc. These have five senses – touch, taste, smell, sight and hearing, i.e. they have a body, mouth, nose, eyes and ears. These are called five-sensed beings.

   Among the five-sensed beings is the first level of animal life where there is no sense of time, viz., what is past, what is present and what is future. These are given 5 life-units on the life-axis.
8. After animal life, the next stage of spiritual evolution is the human person, which has a sense of time or a high degree of coherence in addition to the five senses above. This class is very broad, and thus, for example, a criminal would receive a lower score along the life-axis than a humanitarian.

For an average human, the base score of 100 life-units has been agreed so that a criminal may score only 10 life-units.

Some form of ascending 'scores' can be connected to the idea of the spiritual advancement of individuals:

1. At the first stage are the saints (*Sadhu*) who are supposed to walk on the spiritual path with single-mindedness.
2. Those at the second stage are the spiritual teachers (*Upadhayaya*) who have experienced the truth.
3. The third are the spiritual masters (*Acharya*) who practice what they preach, being the true masters.
4. The fourth category (*Arihanta*) is that of perfect living beings who have conquered their inner enemies. The nominal life-units for categories 1-4 are $10^3$,$10^5$,$10^{10}$,$10^{100}$ respectively.
5. Those in the final category are pure or liberated souls (*Siddha*), which are a form of absolute energy. The score for the liberated soul is at the point of infinity, as it has no impurity (not even a body).

Early practitioners of Jainism emphasized that teachers from other religious or philosophical traditions could reach the higher states of consciousness. This is a reflection of the pluralist logic at the heart of Jain thought processes.

Most Jains today might not accept the propositions above literally and the quantifications or 'scores' above are intended as guidelines to illustrate the relationship between different states of being and consciousness.

## The Four *Gati* or 'States of Existence'

Every living being possesses varying degrees of sensitivity conferred by its mental state. We describe the four main directions, which the mental state can take. The state with the highest agonizing point is the hellish state. The extreme state of pleasure is called the heavenly state. This is a hedonistic pleasure but does not correspond to the state of bliss. The state where the living being does not know what is tomorrow or did not know what was yesterday is the animalistic state. The state of the equilibrium point between the pleasure and the pain is the average human state.

Every living being is capable of taking the above four directions of these mental states, namely:

Hellish State
Heavenly State
Human State
Animalistic State

These four conditions, or *Gati,* are also referred to as karmic destinies. They are represented symbolically in the form of a *Svastika,* the central point of which represents the capacities of the mind (see Image 3). The Svastika is a symbol of the source of life in Jainism, as it is in many other religions including Hinduism and Tibet's indigenous tradition of Bon (where it is frequently depicted in reverse form). It was misused by the Nazis to represent raw power and the dominance of one human group over others.

The Gati are influenced by the density of the karmic matter, and they should be taken into account while placing a living being on the life-axis of a given species. The literal interpretations of these states correspond to the four states of existence:

1. *Naraki* – Hell Being
2. *Deva* – Heavenly Being
3. *Tiryancha* or *Tiryanka* – (non-human) Animal or Plant Life
4. *Manusya* – Human Being

At the central point of the Svastika the axis of rotation passes through different lives. Within the Jain community, or *Sangha,* the four arcs of the Svastika are also used to represent the four human conditions: male ascetics; female ascetics; lay women and lay men.

Our approach to the Gati is influenced by that of

Kundakunda, the Digambara monk and thinker of the second century CE. He taught that 'the self by its own thought activity creates for itself the four forms of these beings'. This means that the greater the level of self-awareness, the greater the capacity to control one's destiny – and eventually to realize the higher self.

At all times, it must be kept in mind that humans – *Manusya* – only have partial understanding of the universe and the laws of nature. Therefore, we should be constantly aware of other possibilities that will lead us to amend or revise our conclusions. For example, the existence of spiritually advanced life in other parts of the universe is far from discounted by Jains and can hold human arrogance in check. Therefore Jain science is not a closed or exclusive system and its conclusions are not absolute truths but stepping stones on the way to the truth.

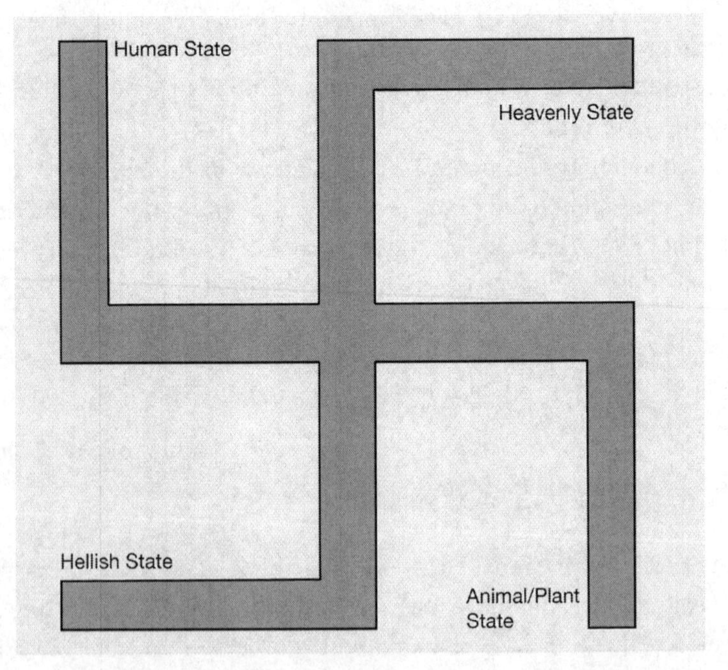


Image 3: *The svastika is a life-affirming symbol for Jains*

## Chapter 5

# Truth 3: Cycles of Birth, Death and Rebirth

Axiom: *'The karmic bondage leads the soul through the four possible states of existence (cycles).'*

### The Axiom

The second Truth concentrated on the static situation of living beings through one life-span rather than the dynamic situation of various life cycles. From this, the question arises as to whether there exists a cycle of birth and death. The third Truth or Axiom of Jain Science assumes that there is such a cycle. On death the soul is set free of the physical body and is thus ready to move on under its own propulsion.

This view of death points towards an important difference between Jain philosophy and the prevalent forms of western scientific humanism. From the western humanistic perspective, death tends to be viewed as an extinguishing of consciousness and a literal 'end' for those who die. The popular expression that 'you've only got one life, so make the most of it' derives from this stance, as does the more advanced position that each individual, and hence humankind collectively, has 'only one chance' and so should make life on Earth as bearable and agreeable as possible for present as well as future generations.

From the Jain perspective, both scientific and spiritual, physical death is not an end in itself as much as a point of transition from one embodiment or incarnation to another. This is because the Jiva or 'soul' is the element that is truly 'alive', whereas the physical body is Ajiva, or purely material in form. The physical death of the body marks its separation from the Jiva, which is its animating principle giving it distinctive qualities of consciousness and perception. These are the qualities, some obvious, others more subtle, that we have come to associate with both 'mind' and 'spirit': in Jainism, these are continuous with the body rather than separate from it. The animating characteristics of the body – namely the properties of the Jiva – arise from karmic inheritance, augmented by the accumulation (or diminution) of karma during an individual life span.

From Axiom 2, it is therefore quite clear that the amount of karmic matter will be responsible for the next placement on the life-axis. The karmic matter is 'carried over' into a new incarnation in much the same way as genetic inheritance affects physical (and many mental) characteristics. From a purely Jain perspective, indeed, genetic inheritance is but another aspect of the karmic process. Death and its counterpart rebirth are not ends and beginnings in themselves, but points on the samsaric cycle, which is widely depicted as a karmic wheel. The process of Samsara ends when the Jiva achieves liberation from karmic entanglement.

For the spiritual and scientific seeker alike, two main questions arise from the second Axiom, which the third aims to address:

1. What is meaningfully transported from one life to the next?
2. What form of science can explain or even conceive of such transportation?

## Karmic Components

To answer the above questions, we assume that the karmic matter gets differentiated into eight specific types by the activities of the contaminated soul. Two of these types, the 'bliss-defiling' and the 'feeling-producing' karmas, are in turn sub-divided into two parts, meaning that there are really ten types of karmic accumulation that we need to consider. Each of the karmic types is an agglomeration of specific types of karmic particles, accumulated (or shed) according to behavioral patterns in present and previous embodiments. We shall call these types the Karmic Components: Jains at times also refer to each type as a 'karma', or in the plural as 'karmas'.

Within the Karmic Components, there are two main types, known to Jains as *Ghatiya* and *Aghatiya* karmas. Ghatiya karmas are, at one level, more 'serious' than their Aghatiya counterparts because they are related to spiritual and mental delusions (of power or importance, for example) which attract the heaviest and most damaging accumulations of karmic matter. This is why they are often described as 'negative', 'destructive' or 'heavy' karmas, weighing down the Jiva with material attachment and separation from the inner self. However, they are also the karmas that can most easily be destroyed or can decay within an individual life-span through awareness and positive action – and, eventually, a life directed by spiritual concerns. In other words, consciousness of these karmic types (and the imperative to transcend them: the 'freedom-longing catalyst') can lead to the most creative acts and the greatest spiritual advances.

Aghatiya karmas, by contrast, are often identified with lightness and positive qualities. More properly, they should be considered as neutral forces. Their consequences are less potentially severe than Ghatiya karmas, but they are also less malleable. For example, it is an Aghatiya that determines the length of the life-span – and hence the opportunities for reducing or accumulating karmic influences during the allotted period. All Aghatiyas are formed through the attraction of karmic particles during one life-span and directly determine the specific qualities and conditions of the next incarnation in the cycle. The influence of the Ghatiya karmas is less direct, but ultimately more powerful and subtle.

For the purposes of this study, we shall call the Ghatiya karmas *Primary Components* and the Aghatiya karmas *Secondary Components*. They may be listed as follows:

## KARMA (8 KARMIC COMPONENTS)

A. Ghatiya – Primary

a. *Mohaniya* – Bliss-defiling; deluding (karma)

(a1) *Darshana-Mohaniya* – Insight-deluding

(a2) *Charitra-Mohaniya* – Conduct-deluding

b. *Virya-antaraya* – Energy-obstructing

c. *Jnana-avaraniya* – Knowledge-obscuring

d. *Darshana-avaraniya* – Perception-obscuring

B. Aghatiya – Secondary

e. *Vedaniya* – Feeling-producing

(e1) *Sata-vedaniya* – Pleasure-producing

(e2) *Asata-vedaniya* – Pain-producing

f. *Nama* – Body-producing

g. *Ayu* – Longevity-determining

h. *Gotra* – Environment-determining (includes social conditions and status at birth)

We will view the karmic components as, in essence, negative forces, despite the positive and creative opportunities that might arise from them. The source of the karmic components is karmic matter and the perverted or distorted energy element of the soul. We have already described the way Jain teachings classify the four basic elements of the Jiva or soul: bliss, energy, knowledge and perception and the freedom-longing catalyst that is one of its innate properties. On the positive scale, we can identify infinite bliss, energy, knowledge and perception. Underlying these elements is the strong freedom-longing catalyst. On the negative side corresponding to bliss, we have a karmic component which defiles this element. This component can be described as the Bliss-Defiling karmic component; we shall call it the 'a-component'. The a-component has an Insight-Defiling sub-component ($a^1$) and a Conduct-Defiling sub-component ($a^2$)

which we shall call the ($a^1$) sub-component and the ($a^2$) sub-component respectively. Recall that the Defiling-Component changes the overall structure of the soul; that is, the process leads to a very fundamental transformation of its elements, e.g. change of personality under intoxication. Similarly, the second negative component obstructs the operation of the energy element; this we will call the Energy-Obstructing Karmic Component (b) and denote by b-component. This makes the soul not only work with restricted energy but it also becomes an accomplice in the process of karmic fusion with the existing karmic matter as well as in karmic decay. Similarly, we have the third and fourth karmic components, the Knowledge-Obscuring Component (c) and the Perception-Obscuring Component (d) which we will write as the c-component and the d-component respectively. Note again that these last two components only obscure the two soul elements and do not defile the soul.

The four Primary Components (a-d) are in constant operation and affect every area of the Jiva's experience. They also fluctuate and are subject to modification. The Secondary Components (e-h) are more static and although their effect on the embodiment of the Jiva is immediate and direct, they attack the freedom-longing catalyst only indirectly. These components will be classified as the e-component, f-component, g-component and h-component respectively. They react only slowly to the process of fusion and decay at a particular moment, except at the time just before the beginning of the next life-cycle. Although all karmic components operate independently, the defiling a-component plays a central role since it defiles the soul and allows other components to operate. In fact, the b-component is influenced by the existence of this defiling process. We may compare these karmic energy levels to those of the electrons in the inner and outer shells of an atom.

## What Gets Transported?

As described above, the four secondary karmic components are responsible for various aspects of the next incarnation. In particular, the body-karmic component is said to generate two 'subtle bodies' or *Sharira*, underlying the manifest physical body:

1. The *Taijasa Sharira* or *luminous capsule,* which maintains the vital functions (temperature, etc.) of the organism.
2. The *Karmana Sharira,* or *karmic body,* constituting the sum total of the karmic matter associated with the Jiva at a given time. The existence of the Karmana Sharira is important to an understanding of the theory of reincarnation because it constitutes the vehicle by which a Jiva moves under its own power from one incarnation to the next.[1]

At the moment of death, the Body-Producing karmic component (f-component) has pre-programmed the particular conditions of the coming embodiment. This information is carried in the karmic body. At death the Jiva is released from its physical body and is said to travel in a straight line almost instantaneously to the destination which its accompanying karmic matter has predetermined.[2]

The transported material – the Taijasa Sharira or 'luminous capsule' – can be viewed as a hermetically sealed unit containing the karmic body and the soul, thereby preventing any further fluctuations, including the shedding of karmons. In spite of the propulsion caused by the soul energy at the time of death it cannot travel too far before it enters a physical body in an egg or womb. The stationary medium ensures that, unless it is being liberated, the Jiva tends to remain within a broad ecological and temporal context in its next embodiment.

## The *Dravya* or 'Six Existents'

Jain teachings have transmitted a series of 'laws of nature' governing (among other things) the interaction between soul and karmons, the next embodiment of the Jiva, and the liberation of the soul (Moksha). According to Jain Science, the universe is comprised of six 'existents' or *Dravya,* some of which will already be familiar:

1. *Jiva* – Soul
2. *Pudgala* – Matter
3. *Akasha* – Space

3.1. *Loka Akasha* (also *Kashaloka* or *Lokakasha*) – Occupied space; inhabited universe

3.2. *Aloka Akasha* – Unoccupied space; uninhabited universe

4. *Adharma* – Dynamic medium
5. *Dharma* – Stationary medium
6. *Kala* – Time

In contrast with standard physics where one deals with matter in time and space coordinate systems, in Jain Science it is the soul which is to be studied in terms of time, space and matter. These all are regarded as 'substances' and will be considered below in the order most compatible with scientific method:

### *1. Space: Akasha*

Jain space is sub-divided into two types: the first, which is occupied by the other five existents, and the second which is empty. We shall refer to these subdivisions as *occupied* and *unoccupied* space respectively.

Occupied space is equivalent to the manifest universe in which all the other five existents are confined. The inherent quality of occupied space is its ability to provide a 'home' for the other five existents and it is divisible into infinitesimally small *space points – Pradesh* in Jain teachings – which have dimension but cannot be further sub-divided.[3] The idea that the occupied universe is bounded is implicit in this formulation. Furthermore, the boundary between occupied space and unoccupied space is considered important, as we shall see later.

### *2 and 3. Dynamic and Stationary Media: Adharma and Dharma*

The 'Dynamic Medium' allows interaction/motion to take place between/within soul and matter, whereas 'the Stationary Medium' allows equilibrium/stability between/within soul and matter. The principle of dynamic and stationary media expresses the principle of continuity and change that is important to every aspect of Jain philosophy and culture, affecting ethical and cosmological principles in equal measure. From the Jain standpoint, each is incomplete without the other. Continuity becomes sterile routine without the ability to modify or transform, whereas change becomes random and vacuous without secure foundations.

The usual analogy is that the dynamic medium is like water allowing the movement of a fish whereas the stationary medium is like the shade of a tree which allows travellers to rest. Thus soul/matter has the inherent quality to 'go' or 'stop', but these two media make these operations possible. In general, the 'Go-mode' includes developing, interacting, moving, etc. and the 'Stop-mode' is the opposite.

The two media are non-atomic, inactive, formless and continuous. These co-exist and we can regard dynamic and stationary media as secondary and tertiary space respectively. The logic behind these two media is elegantly described as follows by the Indologist A.L. Basham (1958, p. 76), in a survey of Jain and Buddhist cosmology. We have adapted the quotation below by substituting our preferred terminology:

*The existence of dynamic medium as a secondary space is proved to the Jain's satisfaction from the fact of motion; this must be caused by something; it cannot be due to time or the atoms, since they have no spatial extension, and that which is spaceless cannot give rise to movement in space; it cannot be due to the soul, since souls do not fill the whole universe, but motion is possible everywhere; it cannot be due to space, for space extends even beyond the universe, and if space was the basis of motion the bounds of the universe would fluctuate, which they do not; therefore motion must be caused by some other substance which does not extend beyond the universe, but pervades the whole of it; this is what is called dynamic medium. The existence of 'stationary medium' is proved by similar arguments.*

The first four existents – soul, matter, space and time – do not themselves undergo any changes due to the two media, but they function in so far as soul and matter in either 'Go-mode' through space or 'Stop-mode' in space. Thus in particular, the dynamic medium allows karmic fusion/fission whereas the stationary medium allows the state of karmic bondage. Further, the dynamic medium will allow the soul to travel to the next embodiment whereas the stationary medium will allow it to be planted in a womb.

We have regarded the two existents as media for motion and rest but these can be viewed as two forces: *Dynamic* and *Stationary.* These two polarities operate on the levels of both *Jiva* and *Ajiva* – living and non-living or purely material. It is worth noting that the word 'Dharma' is also the word for the workings of the universe, natural law or unchanging constants within nature, with which the precepts and vows of the Jains seek alignment.

### 4. Time: *Kala*

Time is not affected by the other existents. Jains believe that time is digital, i.e. consists of an infinite series of discrete time points.

Time as an existent has no beginning or end. This corresponds with the Jain view of energy as an inherent force within the universe that – in terms familiar to modern physics – can 'neither be created nor destroyed'. The universe is also viewed as moving through innumerable upward and downward cycles of time, constantly renewing itself and therefore 'eternal'.

### 5. Matter: Pudgala

The word 'Pudgala' will be rendered here as 'matter', which is the most usual translation. Yet in Jain Science the word has always encompassed 'physical energy'. The word is formed from the two words *pum* (joining) and *gala* (breaking). This gives central importance to the formation and destruction of matter. 'Destruction', in this context, implies converting matter into energy and energy into matter. In modern physics, the term 'mass energy' broadly covers this concept, but it is important to remember that Jainism focuses on 'matter-energy' and the complex and intimate relationship between them. The concept of 'joining' and 'breaking' also expresses the workings of continuity and change.

Matter is finally composed of what may be described by Jains as the *Paramanu*. This is at best translated as *the ultimate particle* (UP). The Paramanu are the smallest indivisible particles, which can be aggregated in many different ways so that they produce every form of organic and inorganic matter but exclude the Jiva or soul itself. Karmic matter on a contaminated soul is a fine matter which has an *infinite number of karmons.*

The luminous capsule or Taijasa Sharira is translated by some writers as 'magnetic body' or 'electrical body'. It is also claimed that it is a body of luminous matter and is a necessary link between the other two bodies of the soul, the karmic body and the physical or gross body. A link of this kind is needed because the matter of the karmic body is too fine and that of the physical body too gross to allow any direct or immediate interaction between them. C.R. Jain, for example, puts forward this explanation in his book *The Practical Dharma,* first published in 1929 and reissued as *Fundamentals of Jainism* in 1974. This is a good example from the first half of the twentieth century of a movement by Jain intellectuals to reclaim the inner teachings of their faith and at the same time engage with the modern worlds of science, politics and law.

As the soul has its characteristics of life including bliss, energy, knowledge and perception elements, Pudgala has its characteristics of lifelessness, touch, flavor, smell and color. The important principle is that each quality produced by elementary particles undergoes constant changes of mode along its respective continuum. Thus, matter and energy may be regarded from the Jain standpoint as one and the same thing, i.e. sound, light, heat, etc. are types of matter but their *mode* is energy. These Jain concepts of matter and energy do not seem to include all the concepts of modern physics but nevertheless, these are compatible (see Chapter 10). On the other hand, Jain Science explains the phenomenon of Mind over Matter. It shows how finer karmic matter from karmons and the soul are interrelated.

## *6. Soul: Jiva*

Occupied space contains an infinite number of Jiva or souls. Each Jiva has an uncountable number of space points but exists within the physical limits of its current corporeal shape. Liberated souls are all distinct and are not under constraints of time, dynamic or stationary forces and are on the highest point of the boundary between occupied space and unoccupied space. The highest point on the boundary is perhaps similar to a black hole in the sense that the standard laws of physics are not applicable in this context.

When all karmic matter, even the finest, is removed, the soul will move to this highest point. The soul now attains infinite bliss, energy, knowledge and perception, the state of liberation (from karmic restraint) known as Moksha.

Note that in Jain teachings, the mind is regarded as the sixth sense, made up of matter which acts as a processor of input from the five senses and it should not be confused with consciousness – the knowledge and perception elements.

## Jain 'Particle Physics'

Pudgala or 'matter-energy' is held to have one of five colors, one of five flavors, one of two odors and one each of the four (sometimes five) pairs of touches:

Colors (5): Black, red, yellow, white, blue
Flavors (5): Sweet, bitter, pungent, acidic, astringent
Odor(s): 'Good' smell, 'bad' smell
Touches (4 or 5 pairs): Hot/cold, wet/dry, (smooth/rough), hard/soft, light/heavy

The *Paramanu* or Ultimate Particle (UP) has the following properties:

One of the five colors
One of the five flavors
One of the two types of odor
One of the four usual pairs of touches – either *wetness/dryness* or *hot/cold*.

Through these formulae, Jain Science traditionally arrives at 200 distinctive 'primary UPs'. It is shown that wetness and dryness possess varying intensities which are integers. These combine together to produce *composite* bodies. The fundamental condition is that the UPs in combination must possess more than one unit intensity of dryness or wetness. These cannot combine if the intensity is only one unit. Karmons (karmic particles) are some of the finest particles formed from UPs. In Jain science, sub-atomic (including karmic) particles are referred to by the generic term *Anu*. Particle groupings ('variforms' for physicists) are known as *Vargana*.

## Some Practical Implications

It is apparent that karmic matter plays a central role in shaping one's next incarnation along the life-axis. Hence, an average human being indulging in criminal activities might end up in the next life as a snake because of the heavy karmic matter accumulated. On the other hand, that same average human being can, after expiation of his heavy karmic matter, ascend the spiritual ladder. The cycle subsequently continues: for example, the one who has become a snake could, after reducing karmic matter, 'ascend' to the level of human being in his second cycle, with the spiritual intelligence and responsibilities that this state demands.

There are five caveats associated with the examples above:

1. The intention behind the 'criminal' activities might be as important, or more important, than the activities themselves.
2. A spiritual teacher need not always have benign motives. He or she might be motivated by power or material gain, which will attract the heaviest forms of karmic matter. (This is why Jains are asked to choose their teachers with great care and remember that each person is ultimately his or her own guru.)
3. The snake has a place in nature in the universe and is an embodiment of the Jiva. It should not therefore be regarded as a creature lacking worth, although it is held to have less intrinsic capacity for spiritual development than an average human being.
4. It is possible in some circumstances for a snake to shed its heavy karmic matter (much as it sheds its skin).
5. The term 'average human being' is not a scientific term,

> but an approximation. Each individual has unique properties as well as shared characteristics with the rest of humanity and there are a wide variety of cultural differences to keep in mind, as well as differences of conditions and values between human societies and different historical epochs.

Jain teachings, again like cutting-edge science today, associate the pursuit of knowledge with constant awareness of caveats and multiple possibilities.

In view of Axiom 2, one can end the cycles only through the human state for which the karmic density is comparatively lower than for any other form of life. Axiom 3 (or the Third Noble Truth) teaches us that the processes associated with karmic bondage keep us enmeshed in the cycles of existence from one incarnation to another. It shows us that by conscious thought and careful action we are able to influence the nature and quality of these incarnations: we can do this by the choices and decisions we make in our present 'form' – that is to say the present embodiment of the Jiva that animates us.

Therefore it follows that we should care about present and future social (and environmental) conditions *because* we have many lives rather than only one. Or, to put it more accurately, we are part of a far larger life process than our present (embodied) selves.

The methods of removing all the karmic matter in the human state (i.e. finally cutting the bonds that imprison the soul), will be explored in the next chapter.

However, when one's soul is liberated from the cycle of rebirth, Jains believe that immediately another soul from a low form of life shoots higher; this leads in turn to souls in lower forms moving higher.

Therefore, in liberating ourselves we are helping a lower life form to rise up on the life-axis.

This chain-like progression is an interesting concept and appears to be unique to Jain teachings.

# Chapter 6

# Truth 4A: Practical Karmic Fusion

Axiom: *'Karmic fusion is due to perverted or distorted views, non-restraint, carelessness, passions and activities.'*

## The Axiom

We have seen in previous chapters that the density of karmic matter makes the difference between various species. At the human level the density is small and so the spiritual awareness of human beings has a correspondingly high potential. However, to realize the full power of the Jiva or soul, it is important to remove all vestigial influences of karmic matter from it. Before we try to find how this can be achieved at the human level, it is important to understand how karmic fusion takes place in practice. This enables us to apply in practical terms the more abstract themes developed above, so that Jain Science makes the transition from its theoretical to its applied form.

The karmic force field is set up by a series of *Activities,* known as *Yoga* in Jainism. These encompass all the activities of the mind and body, including speech, which gives expression to desires both superficial and profound. Karmic fusion takes place as a result of the volitional activities of the individual, i.e. the exercise of one's own will. Volition is known to Jains as *Bhava.*

It is important to note that activities *in themselves* cannot set up a karmic force field: a new-born child, for example, acts but has not acquired a sense of volition. When Bhava is acquired through the development of consciousness, volitional activities are performed, and in this way the karmons are attracted and fused.[1]

Jain Science recognizes five 'karmic agents':

1. *Mithyadarshana* – Perverted or distorted views

2. *Avirati* – Non-restraint
3. *Pramada* – Carelessness
4. *Kashaya* – Passions
5. *Yoga* – Activities

Kashaya are in turn divided into four categories or sub-agents:

4.1. *Krodha* – Anger
4.2. *Mana* – Pride
4.3. *Maya* – Deceit
4.4. *Lobha* – Greed

*No-Kashaya* are 'subsidiary' Kashaya, into which categories 1-4 are grouped according to type. There are two categories:

1. *Raga* – Attachment
2. *Dvesha* – Aversion

The five Karmic Agents influence the acquisition and effects of karmic matter. They work on and undermine the four Guna or 'soul-elements', the ingredients of the Jiva:

1. *Jnana* – Knowledge
2. *Darshana* – Perception
3. *Sukha* – Bliss
4. *Virya* – Energy

*Mithyadarshana* – perverted or distorted views – means false notions regarding the nature of the soul or understanding of the true self. Ideologies based on hatred, domination, exploitation and self-righteousness all stem ultimately from this type of false understanding. In Jain terms, the Knowledge and Perception Guna are obscured.

*Avirati* or Non-restraint implies absence of self-control which may lead to involuntary evil deeds. Thus, the Bliss Guna is defiled.

*Pramada* or Carelessness implies general inertia in working towards enlightenment. Thus the Energy Guna is obstructed.

Jain Yoga refers to general activities of the body and mind – as well as speech, which is considered to be such a powerful force (for good or ill) that it becomes a *yoga* in its own right. Positive Yoga (= sacred activities) leads to light karmic matter whereas Negative Yoga (= harmful activities) leads to heavy karmic matter. The last of the agents responsible for karmic fusion is Passion. This is the main agent for fusion, and it influences all four 'soul elements'.

This definition of *Yoga* as all types of action differs radically from the physical and spiritual activities (e.g. Hatha Yoga and Raja Yoga) that have arisen from the Hindu tradition. In the Vedic Dharma, the term 'yoga' is synonymous with spiritual union of human and divine qualities – and activities that lead to that union. For Jains, yoga means all actions, whether beneficial or harmful, sacred or secular.

## How Karmic Components Work

The Insight Deluding Component gives rise to false views including extremism and an inability to discriminate between what is proper and what is improper. The Conduct Deluding Component generates passions and sentiments which delude right conduct. These two sub-components act simultaneously to create a state of spiritual blockage. The Knowledge Obscuring Component obstructs knowledge in five ways. It obstructs:

1. The function of the senses and mind
2. Logical ability
3. Clairvoyant insight

4. Intuition or 'mind-reading' ability
5. The achievement of omniscience

These are all properties of the undiluted Jiva or liberated soul. Through Moksha, the soul emerges into full consciousness of these abilities. The spiritual path, which is simultaneously a quest for knowledge, is a process of developing as many of them as far as possible, although full omniscience is not possible without full enlightenment. All of these qualities of the soul are interconnected and dependent on each other. Clairvoyance and mind-reading intuition are considered by Jains to be powerful spiritual properties, to be developed and used with care.

The Perception Obscuring Component obstructs perception by means of the eyes and other senses, including the obstruction of intuitive and clairvoyant powers.

The Bliss Defiling Component (Insight Deluding and Conduct Deluding) limits the energy of the soul and activities of the body, mind and speech. In this way, it sows confusion and generates desires that trigger new karmic components into action. Its effects are the spiritual counterpart to alcoholic or drug-induced intoxication.

We may now summarize the secondary set of karmic components, which define identity, status and circumstance and this can have profound effects on the ability of a Jiva to develop within a given embodiment:

1. The Feeling Producing Component characterizes mental state.
2. The Body Producing Component determines the type of species, sex and color.
3. The Longevity Component determines longevity in the next birth.
4. The Environmental Component determines the level of circumstances conducive to the pursual of spiritual life.

## Karmic Dynamics in Practice

We now give the details of the karmic dynamics in practice. Let x be the number of karmons involved in fusion due to a volitional activity. Note that the new karmic matter remains dormant for some time before the emission begins.

The precise number of karmons, x, in fusion depends upon the *degree of volition* with which the activity was carried out. The distribution of x over the different karmic components depends on the *type of activity,* i.e. the type of activity determines the specific karmic component taken up by the undifferentiated karmons.

The time to decay and the corresponding potential strength of each component is fixed by the *degree of passions* with which the activity takes place. Once the karmon has had its effect, it is emitted from the soul, returning to an undifferentiated state and thus to the infinite pool of free karmons.[2] Note that the time of activation, duration of emission and the strength of each karmic component can vary considerably. Also, it is possible to enforce premature karmic decay, or suppression of the effects of karmons, through practical means which will be considered below.

Kashaya, or Passion, is the main agent for karmic fusion. Its four 'sub-agents' – Anger, Pride, Deceit and Greed – can be described as the four *principal* passions. Gluttony and covetousness are both regarded as expressions of greed. In essence, the Kashaya are negative human qualities that obstruct spiritual development and self-awareness. They are recognized as such by all the spiritual pathways of humankind and can therefore be viewed as aspects of a universal truth. However, unlike Christianity for example, Jainism does not have a concept of 'deadly sins', because the Jain path involves overcoming the cycle of birth, death and rebirth, and passing through many 'existences' in the process.

The attraction of karmons is stronger on greed and deceit but weaker on anger and pride. However, both can occur simultaneously. Given a particular situation, the activities of body, mind and speech occur, activating the karmic field. Karmons are picked up and then attracted or repulsed by the Four Passions. The incoming karmons go through the process of fusion to the existing karmic matter (underlining the Four Passions for simplicity) through the energy element of the soul; this is one's personal reaction in view of the existing karmic matter.

The karmons are then assigned a function depending on subsequent volitional activity, i.e. a righteous action will lead to light karmic matter being added. An unrighteous action leads to heavy karmic matter being added. This means a *weaker* or *stronger* form of karmic fusion respectively.

In classifying actions as 'righteous' and 'unrighteous', Jains are constantly aware of the intention behind the act. For example, an apparently charitable or humanitarian act can be sullied by an impure motive such as self-aggrandizement, desire for power or attempts forcibly to impose a religion or ideology on others. Therefore, intention counts as an activity or yoga in its own right.

It should also be noted that Anger and Pride are grouped as Raga or Attachment whereas Deceit and Greed are grouped as Dvesha or Aversion. These two subsidiary Kashaya reflect the broad emotional states into which the categories of passions can fall.

## Degrees of Kashaya (Passions)

We may now illustrate the strength of the main Kashaya or passions – anger (*krodha*), pride (*mana*), deceit (*maya*) and greed (*lobha*) – assigning to them five degrees: 0, 1, 2, 3, 4. Of course, these imply the proportional density of fusions of karmons, that is, the higher the degree, the larger is the fusion, the longer is its time to decay, and the stronger is the karmic force.

The degrees of anger, pride, deceit and greed of 0, 1, 2, 3, and

4 can be illustrated through the following metaphors:

### *Anger:*

Degree 1 is like a line drawn with a stick on water which almost instantaneously passes away.

Degree 2 is like a line drawn on a beach which the tide washes away.

Degree 3 is like a ditch dug in a sandy soil which, after one year's weather, silts up.

Degree 4, the most serious or 'worst of all', is like a deep crack in a mountain side which will remain until the end of time.

Degree Zero, by contrast, implies serenity and tolerance.

### *Pride:*

Degree 1 is like a twig which is pliable and easily bent.

Degree 2 is like a young branch of a tree which can be bent by a storm.

Degree 3 is like beams of wood cut from a mature tree which may only be bent by being oiled and heated.

Degree 4 outdoes any analogy taken from a tree, being as unbending as a lump of granite.

Degree Zero implies humility.

### *Deceit (or 'Crookedness'):*

Degree 1 deceit can be 'straightened' as one could straighten a stalk of wheat bent by the wind.

Degree 2 is like the edge of a lawn which has been badly cut and requires much work to straighten it.

Degree 3 is like a crooked tooth which cannot be straightened after it has been left unchecked for a prolonged period.

Degree 4 is akin to a knot in a tree.

Degree Zero indicates straightforwardness, honesty or candor.

### *Greed:*

(In Jain tradition, greed is said to alter the color of the human heart.)

Degree 1 will stain the heart yellow like a water-based paint that can be easily washed away.

Degree 2 means that the heart will be soiled like cooking pans full of fat which can only be cleaned with great labor.

Degree 3 produces a spiritual stain like the mark left by oil on clothing which is only removed after much dry-cleaning.

Degree 4 is like a permanent dye which cannot be removed.

Degree Zero implies contentment and a compassionate attitude.

These degrees can be related to the lengths of the time for which their effects are held to last (for a detailed analysis of this phenomenon, see Glasenapp, 1942):

Degree 4 of a major passion is of lifelong duration.

Degree 3 of a major passion is of one year duration.

Degree 2 of a major passion lasts for 4 months.

Degree 1 of a major passion is the level called smouldering passions and is of a fortnight's duration.

Degree Zero of all major passions corresponds to a higher spiritual state.

Mehta (1939) has included the doctrine of karmic influence and the Kashaya to produce the first comprehensive guide to Jain psychology aimed at a wider, secular readership.

We have examined the four principal Kashaya, namely Anger, Pride, Deceit and Greed. In fact, these four are also responsible for 'quasi-passions' or *sentiments* of nine types: laughter, pleasure, displeasure, sorrow, fear, disgust and sexual cravings outside of committed relationships. Worry and anxiety are included in the category of 'fear'. They are also an aspect of the 'violence to oneself' which is to be discussed in the next chapter.

## Chapter 7

# Truth 4B: Extreme Absorption of Karmons

Axiom: *'Violence to oneself and others results in the formation of the heaviest new karmic matter, whereas helping others towards Moksha with positive non-violence results in the lightest new karmic matter.'*

### The Axiom

The previous chapter revealed the agents that make the karmic flow possible. We have seen that under positive yoga (or 'righteous' activity) the karmons are converted into light karmic matter whereas under negative yoga (or 'un-righteous' activity), the karmons are converted into heavy karmic matter. The emission of that light karmic matter leads to 'good' fruits or benign consequences while the fusion of that heavy karmic matter leads to 'bad' fruits or evil consequences. Light karmic matter may provide a better environment for spiritual progress whereas heavy karmic matter may lead to a lower form of life in future cycles.

In this context, good and bad, positive and negative, are defined in terms of the Dharma or universal law. Negative actions are those which go 'against the grain' of the universe, whereas positive actions work with it. Negative yogas are identified with *Himsa* – harm or violence – and positive actions are identified with *Ahimsa* – harmlessness or non-violence. Ahimsa is the first vow (*vrata*) undertaken by Jains. It is the wellspring of Jain thought and practice. The ideal position is one of total equanimity or an action-less state. This is not easily achieved in a 'normal' human incarnation, although some ascetics come quite close to it, and so the emphasis of lay Jain life is on minimizing harm and choosing positive actions that have benign effects and influences.

The question that now arises is how one gathers the lightest or the heaviest karmic matter. The actions which are responsible for these two extremities of fusion are violence (Himsa) and non-violence (Ahimsa) respectively. In Jainism, the word violence is used broadly to denote anything that does harm. One commits violence to oneself or to others through volitional activities of body, mind and speech, or by urging others to commit violence or by condoning violence committed by others.

Further, the term violence implies any action accompanied by the giving of pain and the heightening of passion. Needless to say, the term includes killing which is reprehensible not only for the suffering of the victims but also for the highest degree of passions which significantly strengthens the killer's karmic bondage.

From Axiom 1, we are aware of the aspiration of all living beings to remove their karmic matter. Helping them towards this objective with dynamic non-violence rather than self-pity is positive non-violence. The intrinsic property of the soul is to 'live and help others to live', for in Jainism the soul or Jiva is literally a 'unit of life'. In other words, the function of every soul is to interact positively with all other souls for the common good. That common good is ultimately defined as spiritual advancement and the liberation of every Jiva. In practice, this means working as far as possible for the good of every 'embodied' (i.e. incarnated) Jiva and developing a social conscience that extends to other species and the environments as well as fellow human beings. The interconnected interests of all forms of life are thereby recognized. Thus the precepts of Jain Science encourage not only the aspirations of a single soul but also, simultaneously, of all souls.

By thinking and living in this way, we increase knowledge and understanding of our true selves. The *Acharanga Sutra* contains the advice that 'You are your own best friend'

(*Acharanga-sutra*, Ch.3, v.125). This is not, as it might initially appear, a spiritual parallel to neo-liberal economics or the spiritual consumerism and 'self-empowerment' associated with the New Age. On the contrary, this injunction expresses the social and ecological dimensions of Jain thought. It is a reminder that we only genuinely 'know' ourselves when we become aware of our shared interests with other living beings. Kindness to all beings is held to be identical to kindness to oneself. At the same time, as we increase our understanding of the true self (i.e. the Jiva), so we also increase our capacity for compassion and consideration for others.

## Some Implications

The idea behind this Axiom is that all living beings are sensitive to pain and no-one desires death. This principle extends logically even to micro-organisms, because they contain Jiva as much as any other embodied being. However, consuming any creature represented on the life-axis necessarily involves killing, so ideally it should be avoided, or rather reduced to a minimum.

For survival one has to consume food and thus we absorb life-units, but the aim is to use the minimum possible number. In general terms, the lower the level of spiritual growth, the lesser will be the total life units. In general, the consumption of life-units of $10^{-3}$ consisting of vegetable life is regarded as acceptable by Jains. However, note that highly concentrated micro-organisms should be avoided since then the life-unit will be above $10^{-3}$. In 'pure' Jain practice, this not only excludes honey and alcohol but also dead flesh as it is an ideal breeding ground for innumerable micro-organisms.[1] It also excludes tissues of certain plants hosting micro-organisms[2] (figs and tomatoes are taken as the symbolic representation of this approach). Vegetables such as onions are avoided by devout Jains, since their life units are considered to be $10^{-2}$. The aim of these guidelines is to minimize Himsa or harm to other life forms, all of which have the potential for spiritual growth. But it must be emphasized once again that the result is not a harsh, prohibitive diet but a rich and varied one. The association that has often been made (even by well-meaning observers) between Jainism and dietary Puritanism is a false one that has been far from helpful to the Jain cause!

Karmic matter that is absorbed through volitional activities affects the individual for a limited period of time only, the length of which depends on the type of action, degree of passion and the motive behind the activity (which counts as a form of prior action). Extreme forms of violence committed under perverted views could have an effect lasting for aeons, whereas if the violence is influenced by any of the Four Passions (Kashaya), then the effect would not be quite so long-lasting. For 'one-sense' life forms, such as micro-organisms, the duration of karmic influence is very much more limited. These beings have few opportunities for the accumulation of heavy karmic matter, but also lack most of the facilities for spiritual growth – until, of course, they reincarnate at a different point on the samsaric cycle. For humans (who are our main concern in this context), the minimum times for karmic decay under Anger, Pride, Deceit and Greed are conventionally taken as two months, one month, a fortnight and less than forty-eight minutes respectively.

We may safely presume that an act of non-violence motivated by, say, greed may have the time decay. However, the maximum decay will be further reduced depending upon the weaker strength of the Four Passions. No karmic matter is absorbed during immobility (the cessation of Jain yoga) and therefore only the remaining karmic matter can be shed.

Implementation of positive non-violence requires full alertness in any action, whether physical, mental or through speech (which is arguably a synthesis of the first two). Mahavira prefixed various discourses to his chief disciple Gautama with the injunction: *'Never to be careless even for a moment'*

(*Uttaradhyayana-sutra,* Ch.6, v.1). This principle is defined in Jainism as having four practical components: Amity, Compassion, Appreciation and Equanimity. These are summarized by Mahavira in the following precept:

> [It is necessary] *to develop a feeling of amity towards all beings, a feeling of appreciation towards the meritorious, a feeling of compassion towards those in misery, and equanimity in instructing those who have lost the true values* (*Tattvartha-sutra,* Ch.7, v.6).

An analogy from modern life could be driving a car towards a given destination: we do well to remember here that a car is a vehicle with tremendous power that needs to be handled with both care and skill. Therefore, the *way* you drive and the degree of care that you take is at least as important as the route that you take. A lapse of less than a second can be crucial.

## 'Accidental' or 'Occupational' Himsa

As we have mentioned, thoughts as well as deeds play an important role in forming heavy and light karmic matter. Thus one should exclude any deed involving 'premeditated violence'. However, such deeds should be contrasted with those which constitute 'accidental/occupational violence'. Thus the number of karmons assimilated by a surgeon even on the death of his patient under an intricate operation is much less than that of a murderer.

Furthermore, the surgeon accumulates only light karmic matter (unless he is incompetent), whereas the murderer always accumulates the heaviest karmic matter.[3] An arable farmer kills insects accidentally in the course of his profession and so he accrues mildly heavy karmic matter. This is because the use of insecticides and pesticides constitutes destruction of life, even though the arable farmer can argue effectively that he cannot 'help' doing this as part of his work. In this context, Jain ethics would favor the use, as far as possible, of non-polluting alternatives to pesticides. This is part of the ecological sensibility at the heart of Jain teachings, including the scientific dimension of Jainism.

In general, the concept of non-violence restricts occupations to those which do not involve premeditated destruction of life above $10^{-2}$ life units. Killing, even when done in the most extreme situation of self-defence – 'defensive-violence' – accrues heavier karmic matter, although this can be mitigated in extreme circumstances, in particular the protection of others. For most individuals such drastic behavior is rarely needed. A famous letter by Mahatma Gandhi to his Jain mentor Raychandbhai together with his answer highlights the spirit (see, for example, Mardia, 1992, 14-15). The aim of Jain ethics is to desist from performing or encouraging others in the premeditated or intentional destruction of souls embodied with two or more senses.

Himsa (violence or harm) in Jain teachings is divided into four sub-sections, including Ahimsa, which as we have seen corresponds to non-violence, harmlessness (in the literal sense) or Zero-Himsa. The other three categories are as follows:

1. *Samkalpaja-himsa:* premeditated violence
2. *Arambhaja-himsa:* accidental/occupational violence
3. *Virodhi-himsa:* defensive

Category 1 corresponds to the example of a murderer who acts 'in cold blood' and/or with the intention of financial and other personal gain (or political gain, in the case of an unjust war). Category 2 corresponds to our example of the farmer using insecticides. Category 3 corresponds to the householder defending his or her children against a violent intruder. It can also correspond to the actions of soldiers or guerrillas defending their country from invasions and occupations motivated by aggressive and exploitative designs. In Jain ethics, a careful distinction is drawn between attack and defense. It is recognized that there are circumstances in which failure to act, even if that means violence, can produce greater harm than the act itself.

To refine our example of the householder, let us suppose she is a mother protecting herself and her children against a violent attacker who has broken into her home. She does so successfully, but only by wounding – perhaps fatally – her assailant with a knife. While this action generates heavy karmic material, the weight of the karmic particles, and their long-term impact, is lessened by the fact that her single act of violence has prevented greater harm. On a grander scale, the conspirators against Hitler in the early 1940s would – had they succeeded – have executed a violent act, but one that might have saved millions of lives. Therefore, the beneficial karmic consequences would eventually have outweighed the negative or heavy impact of the violent action.

In both cases, the motives behind the defensive harm would have assumed a critical importance: had the plotters against Hitler merely wished to establish a variant of Nazi ideology, then the levels of Himsa would greatly increase and the mitigating factors disappear. In Jainism, intention *is* action and can define the karmic consequences of any physical acts that follow. Equally, willful inaction can be as much a form of Himsa as deliberately destructive acts. Often this takes the form of collusion in active Himsa, by which harmful deeds are condoned or met with indifference and denial. A powerful example of this phenomenon was the widespread public indifference to the persecution of the Jews and other minorities in Nazi-occupied Europe, although this was counter-balanced by an equally powerful spirit of resistance. Acceptance of colonialist exploitation (by its supposed beneficiaries) is another example, hence Mahatma Gandhi's campaign of 'non-violent non-co-operation' against the British authorities in India. Also known as *Satyagraha* ('Truth-Struggle'), this campaign was strongly influenced by Jain doctrines. Today, it could be convincingly argued that indifference to growing inequality and the destruction of the environment is a form of passive Himsa. Living non-violently means questioning and reordering our current priorities.

## Cycles of Time

Jains believe that the universe contains various 'worlds' or areas supporting life, including human life. Each of these inhabited worlds goes through an endless series of *Kala* or cycles, half-progressive and half-regressive. Jains call the progressive half-cycles *Utsarpini* and the regressive half-cycles *Avasarpini*. However, their phases are different so that at every moment there is a living Tirthankara somewhere in the universe. These half-cycles are divided into six time-sections. We write **m** for misery and **h** for happiness. For the regressive half-cycle the successive time-sections are:

1. Extremely happy: hhh
2. Happy: hh
3. More happy than unhappy: hmm
4. More unhappy than happy: hmm
5. Unhappy: mm
6. Extremely happy: mmm

We now enter the progressive half-cycle where the successive time-sections are:

7. Extremely unhappy: mmm
8. Unhappy: mm
9. More unhappy than happy: mmh
10. More happy than unhappy: mhh
11. Happy: hh
12. Extremely happy: hhh

It is believed that only during time-sections (3, 4) or (7, 8) is it possible for a Tirthankara or 'perfect being' to emerge. All the twenty-four Tirthankaras of the present half-cycle (regressive) were born during the third (hhm) and fourth (hmm) sections. These combinations of h and m are necessary and sufficient to pursue the course of self-realization. We are now 2609 years (in 2013) into the fifth time-section of 21,000 years so it will be a long while before any more Tirthankaras / perfect beings emerge on this planet. However, the most spiritually developed persons can make contact with a Tirthankara in other worlds as there is always one Tirthankara somewhere in the universe at any instant.[4]

Jambu is assumed to be the last person in the present time-cycle to reach Moksha on Earth: this event took place at around 463 BCE. A verse of the scripture *Kalpasutra* (v. 146) describes when Mahavira instituted the fifth section of the time-cycle mm:

> *The Venerable Ascetic Mahavira instituted two epochs in his capacity of a Maker of an end: the epoch relating to generations, and the epoch relating to psychical condition; in the third generation ended the former epoch, and in the fourth year of his Kevaliship the latter.* (tr. Jacobi, 1884, p. 269)

Hermann Jacobi, the great nineteenth-century German Indologist who translated this verse, interprets it as follows:

> *The meaning of this rather dark passage is according to the commentary that after three generations of disciples (Vira, Sudharman, Gambusvamin) nobody reached Nirvana; and after the fourth year of Mahavira's Kevaliship nobody entered the path which ends in final liberation, so that all persons who before that moment had not advanced in the way to final liberation, will not reach that state though they may obtain the Kevalam by their austerities and exemplary conduct.*

The *Kalpasutra* is the work of a scholar and ascetic named Bhadrabahu, who lived around 300 BCE and is today revered by both the Digambara and Svetambara branches of Jainism. He came from a northern area of the Indian subcontinent that is now a part of Bangladesh, but moved south with a group of devotees to Jaipur. There, he produced his *Niryuktis* or commentaries on Jain principles and the Emperor Chandragupta became one of his leading disciples. Legend tells that he possessed *nimitta jnan,* a power of subtle prescience that comes with great enlightenment. This sensitivity enabled him to tune into hidden patterns in nature and hence predict twelve years of famine in the northern regions. Bhadrabahu's best known work is the *Kalpasutra:* 'Kalpa' literally means a universal law of the universe (akin to Dharma) and 'Sutra' thread, or verse.

In this translation, the word 'kevaliship' means status as a *Kevalin* or one who has achieved *Kevalajnana* (infinite or omniscient knowledge). For most Jains today, the verse implies the entry of humankind – and the 'inhabited universe' of which we are part – into a period where enlightenment in its purest forms is especially difficult or even impossible to attain. There is a possibility that we (that is to say, all embodied souls) might have to wait until this cycle runs its course and a more auspicious one emerges. Yet one of the enduring strengths of the Jain tradition is its ability to transform 'dark' into 'light', finding positive opportunities in seemingly unpromising situations. For the extreme improbability of omniscience is not interpreted as an invitation to apathy or resigned despair, but an invitation to pursue knowledge and ask questions. The more we know and understand about the physical world and the cosmos, the more we can develop as spiritual beings. Thus the division between the scientific and spiritual realms is quite absent from Jain culture. Indeed the Jain sensibility views physics and chemistry, astronomy and cosmology, as some of the most important spiritual disciplines. They help us to connect together fragments of knowledge and so lighten the karmic burden of ignorance. More than that, they bring us closer to an understanding of our true selves and 'That Which Is'.

This Axiom emphasizes the relationship between cause and effect by which the inhabited universe operates. Every action, conscious or otherwise, produces a reaction and even the seemingly smallest acts can have far-reaching and radical consequences. Through this logic, the ethical or moral teachings conventionally associated with spirituality are fused with the discernible laws of physics and cosmology. Scientific and spiritual practitioners are reminded that they are searching for the same truths from different angles or *viewpoints,* because Dharma is as much an ethical system as an overarching scientific law. As we acquire greater knowledge, so we gain the ability – and the responsibility – to act with care and restraint. We become able to live in greater accord with Dharma and so begin the process of *self-conquest*.

Chapter 8

# Truth 4C: The Path to Self-Conquest

Axiom: *'Austerity forms the karmic shield against new karmons as well as setting off the decaying process in the old karmic matter.'*

## The Axiom

From Axioms 4A and 4B, we have understood how the process of karmic inflow works in Jain Science. The process of self-conquest is about transcending karmic influences and so the aim is two-fold:

1. To stop the inflow of new karmons through the karmic shield.
2. To omit the old karmic matter.

If these objectives can be achieved then one will be left with pure Jiva with its full power, namely, its infinite energy, absolute bliss, and perfect knowledge and perception.

It is expected that the full power of the soul can only be achieved by removing the effects of the karmic matter revealed in practical terms by the volitional activities of body, mind and speech. As we have seen, these are the external functions which continuously act and react in the karmic field somewhat after the fashion of a nuclear reactor. Further, there is a process that can be compared to the workings of 'personal karmic computer' associated with the Jiva which keeps an up-to-date record and manages instructions in real time. The question arises as to how one can remove karmic matter as well as stop further karmic fusion.

Rationally, one's slavery to the dictates of one's lower nature should be part of karmic matter since it restrains the soul from having its full power. Hence it is only some form of austerity or restraint that can check the inflow of karmons, i.e. austerity is the only way by which one is able to escape from the constraints of one's physical nature and psyche which are under the continuous influence of the karmic field.

This Axiom promotes austerity as a way of eliminating the five karmic agents of Axiom 4A, viz., perverted views, non-restraint, carelessness, passions and activities. The process of the gradual elimination of these karmic agents, through the practice of austerities, can be presented in 'fourteen purification stages' which we describe below.

Austerities should be understood in a wider context. They imply the control of the senses with extreme alertness while keeping positive non-violence in the forefront. That is, 'Exert yourself according to your capacity', which means that one should not practice austerities to the extent of harming oneself by trying to go beyond one's capabilities. Austerities must never be confused with masochism. They should instead be identified with Careful Action (Iryasamiti) and self-restraint.

## The Purification Axis

**Fourteen Purification Stages of Jain Dharma: 14 = 'highest' stage; 01 = 'lowest' stage**

**14: Static omniscience state**
**13: Dynamic omniscience state**
**12: CSR with eliminated greed**
**11: CSR with suppressed greed**
**10: CSR with subtle greed**
**09: CSR with uniformly mild volition**
**08: CSR with unprecedented volition**
**07: Carelessness-free self-restraint**
**06: EWV with careless SR**
**05: EWV with partial SR**

**04: Non-restrained EWV**
**03: Mixture of deluded and enlightened world views**
**02: Lingering enlightened world view**
**01: Deluded world view**

**Key: EWV = Enlightened World-View; SR = Self-Restraint; CSR = Complete Self-Restraint**

We have already examined the idea of a life-axis. Now we approach the 'upper' portion relevant to human beings who are already spiritually further evolved than other life forms. The purification axis plots the human beings from those with low life-units to those with very high life-units, i.e. from those with maximum karmic density to those who have the lowest karmic density. In other words, it is the upper part of the life-axis which has been extended. This is described in Jain doctrine as the ladder one (i.e. one's Jiva or soul) must climb as one progresses from heavy karmic density to a very low quantity of karmic matter and eventually to liberation.

In this chapter, we undertake a brief survey of the stages on the Jain 'ladder' to liberation. Inevitably, this will involve describing intricate mental processes reflecting many centuries of reflection and speculation by devotees and scholars. The process of self-conquest – the underlying purpose of all Jain practice – can be interpreted in a literal way as described, or as a metaphor for the 'struggle' to transcend the negative or 'lower' aspects of our nature and develop our 'higher' impulses towards compassion, creativity and the pursuit of knowledge: in other words, to evolve as spiritual beings.

The ladder has fourteen rungs which correspond to the fourteen stages of spiritual purification, known as *Gunasthanas* (or *Gunasthana*). We shall call these the *fourteen purification stages.* The higher one is on the ladder, the higher is the degree of purification and the lesser is the karmic matter. In this chapter, we have expressed the Gunasthanas in English translation or as initials, to help the reader navigate this complex area of Jain thinking. For the original Prakrit terms, please see our 'Gunasthana Glossary' at the end of this chapter.

The first point shown on the axis is the first rung applicable to all beings and it is here that the karmic matter is at its densest for human beings. The extent of the karmic matter decreases up the ladder and is zero at the 14th rung. Thus, inversely, we could view the purification axis as through the karmic density axis with fourteen important points on it – the karmic density axis being continuous.

To understand the dynamic process of karmic fission, it should be clearly understood that as karmons are shed, there is an increase in the energy of the Jiva which allows further spiritual growth. It is assumed that future karmic influx will be checked, and there will be a further release of energy and knowledge elements which allows the soul to search for its true nature. Another important point to note is that the effect of the karmic matter is mostly first suppressed rather than altogether eliminated. Furthermore, each stage severely limits karmic fusion and diminishes old karmic matter.

At most of the stages, the degrees of Anger, Pride, Deceit and Greed are reduced gradually, with Anger being the first to be reduced and so on. The five degrees of the Four Kashaya (Passions) have already been described in Chapter 5. However, the overall aim is to eradicate all the five important agents of karmic fusion given in Axiom 4A.

## The First Four Stages of Purification

The first four stages are Deluded World-View (of reality), Lingering Enlightened World-View, Mixture of Deluded and Enlightened World-View and Non-Restrained Enlightened World-View (True-Insight) respectively.

The first rung of the ladder corresponds to all living beings with a Deluded World-View. In the beginning every soul is in this stage of complete ignorance, i.e. it has the Four Passions at the maximum level. However, in view of Axiom 1, every soul strives to release its four elements from karmic matter. This process can be triggered by either internal experience such as remembering past lives or external experiences such as understanding spiritual or scientific teachings.[1] The 'trigger' event is quickly followed by a passage through stages 2 and 3 (defined below) to stage 4 which corresponds to the 'Non-Restrained Enlightened World-View'. This experience is the complete revelation of the true nature of life and the reality of the soul, i.e. True Insight.

This first experience of True Insight lasts only for a few moments and it comes from jamming the insight-deluding karmic component rather than from its elimination. The jammed component will be quickly un-jammed and will assert its influence again. Hence the soul will revert to its extreme, perverted (or distorted) stage with all five karmic agents – Perverted (or distorted) Views, Non-restraint, Carelessness, Passions and Activities – operating with full force.

However, during this 'fall', the soul goes through the third purification stage for a short time where the gross passions remain suppressed but there is no longer True Insight; this stage is described as the stage of Mixture of Enlightened World-View. Below this is stage 2, the Lingering Enlightened World-View stage, in which the fourth degree of the Kashaya (Passions) is reasserted and instantaneously drops the soul down to the first stage again.

In the first transition to the fourth stage, the insight-deluding component is suppressed only, but in subsequent (guaranteed) transitions, of longer durations, there is also partial elimination of this component. After a number of such transitions involving partial elimination-cum-suppression, the soul gets firmly established in the fourth stage to proceed to the fifth stage and beyond as it is described below. The first four stages may be summarized as follows:

1. Deluded World-View: Distorted (or 'Perverted') stage of consciousness
2. Lingering Enlightened World-View
3. Mixture of deluded and Enlightened World-Views
4. Non-Restrained World-View: understanding of true nature of life/reality

At the fourth stage Perverted (or Distorted) Views are removed and equanimity is attained. It is this increase in purity which allows the flash of True Insight to take place. The removal of the fourth degree of the Four Passions leads to increased energy and knowledge elements of the soul, which makes the soul search for true knowledge more vigorously than before. Also it places significantly less emphasis on the manifestation of karmic matter including on one's own body, psychological states seen through the Four Passions, and one's personal possession to which it had formerly identified itself.[2] Thus, a pure and serene state is attained and this affects three main areas of the human personality:

### 1. Attitude and Inner Self

By now, one's serene state encourages an attitude which wishes to address the question 'Who am I?' This attitude exerts the perception element of the soul even further and, through a surge of the energy element not before experienced, further removal of karmic matter takes place. Permanent attainment of True Insight is now possible. This change in consciousness in turn plants a desire to overcome the effect of karmic forces, and this desire leads to a further release of the energy element. All obstructions to insight are thereby prevented from exerting any influence and at that moment the soul experiences a permanent view of reality. Thus identification with the true inner-self, as opposed to 'I', is achieved and so the 'bliss' element of the Jiva is now experienced at a deeper level.

## 2. Behavior and Positive Non-Violence[3]

Being at peace with oneself induces a profound change of behavior and outlook. One becomes aware of the fundamental similarities of all living beings and this feeling of togetherness generates amity towards all and great compassion for the less fortunate. This *compassion* is free from pity and free from any personal ties with a particular being. Due to this realization the soul recognizes that all creatures are potential candidates for liberation. There is an unselfish longing to help other souls towards liberation with equanimity. Positive non-violence causes the evils of exploitative and destructive behavior to be recognized. This aspect of positive non-violence is the practical application of Truth 4B.

## 3. Effects on the Four Passions (Kashaya)

In order to reach the fourth stage, austerities are not mentioned explicitly anywhere, but implicitly it is assumed that they are required since, to be in the fourth purification stage, one has to have all the degrees of the Four Passions down to level three. This cannot be achieved without restraint – and in any case self-restraint is the basis of non-violence (Ahimsa).

The first awakening removes some karmons, leading to a moderate degree of self-control/restraint. In other words, the impulses that lead to anger, deceit and intrigue, overwhelming pride and devouring greed are held in check and, with greater spiritual practice, will eventually fade away.

Moreover, on the perfection of the fourth purification stage, there will be evidence of more tolerance and less anger, more humility and less pride, more straightforwardness and less deceit, more contentment and less greed.

## Stages Five to Eleven

When distorted views are replaced by True Insight, one rises to the fourth stage. At the fifth stage one starts working to achieve even greater restraint; that is, one follows various vows that lead to partial restraint. At the sixth stage, full restraint is accomplished.

The fifth stage is equivalent to the way of life of the ordinary lay man whereas the sixth stage corresponds to following the path of a monk. At stage six, i.e. at the state of full restraint, full discipline and higher vows are achieved. How these various stages are achieved is described below in the next chapter.

At stage seven, one reduces Carelessness to zero, implying also that Anger goes to zero and therefore this stage is called the Carelessness-free Self-Restraint stage. However, some remnants of the Four Passions still persist.

At the eighth, ninth and tenth stages, one tries through meditation to decrease the degrees of Pride, Deceit and Greed to zero for each one. The corresponding stages of attainment in meditation are:

Complete Self-Restraint combined with:

8. Unprecedented Volition
9. Uniformly Mild Volition
10. Subtle Greed
11. Suppressed Greed

These seven stages can therefore be summed up briefly as follows:

| | |
|---|---|
| Enlightened World-View with Partial Self-Restraint: | True Jain Lay man |
| Enlightened World-View with Careless Self-Restraint: | Jain Ascetic |
| Carelessness-free Self-Restraint: | Spiritual Teacher |
| Complete Self-Restraint with Unprecedented Volition: | Spiritual Masters |

| | |
|---|---|
| Complete Self-Restraint with Uniformly Mild Volition: | Advanced Masters |
| Complete Self-Restraint with Subtle Greed: | Advanced Masters |
| Complete Self-Restraint with Suppressed Greed: | Passionless State (i.e. free from Kashaya) |

When in these states, if the Four Passions are suppressed rather than eliminated, then one will only be able to reach the eleventh stage. This is known as the 'Complete Self-Restraint with Suppressed Greed State', from which one will be compelled to move downwards. However, if the Four Passions and their effects are fully eliminated during the trances, so that the degree of greed becomes permanently zero, then one will jump straight from the tenth stage to the twelfth stage – the 'Complete Self-Restraint with Eliminated Greed' state.

## Stages Twelve to Fourteen

At the moment of attaining the twelfth stage, three remaining primary karmic components (other than Deluding Karmic Component) are automatically eliminated. This leads in turn to attainment of the thirteenth stage which is the state of omniscience and will be called the 'Dynamic Omniscience State'.

At this stage only *yoga* (action) governs the continuing functioning of the body. These activities do not, however, lead to new karmons. Also, secondary karmic components of the omniscient being gradually fall off until eventually none of them remain. In the final moments, the body is in a state of total immobility – this state is the 'Static Omniscience' stage and is the fourteenth stage.[4] This state lasts only for at the most forty-eight minutes prior to Moksha.

At the moment when physical death occurs the Jiva or soul, completely and forever freed from the cycle of rebirth, attains

Moksha. Thus:

> Complete Self-Restraint with Eliminated Greed
> Dynamic Omniscience State: Tirthankara
> Static Omniscience State: Towards Moksha

Note that stage four is the attainment of 'Non-restrained Enlightened World-View'; stage five is attainment of the household state of lower vows; stage six is the attainment of a saintly level of higher vows; stage seven is very much like the state of a spiritual teacher; stages eight to ten are those of spiritual masters; stages twelve and thirteen are stages of Tirthankara / Dynamic Omniscience. The fourteenth stage is the state of omniscience at the instant prior to Moksha.

This ancient model of self-conquest might appear to lack relevance today for those who wish to understand Jain principles. Much of the terminology, after all, is arcane and is not uppermost in the minds of Jain practitioners. However, it is part of their cultural inheritance and as such influences, albeit indirectly, their view of life and spiritual endeavors. The Gunasthanas tell us much about the culture or *mentality* of Jainness. They demonstrate the scientific and psychological approach taken by Jain practitioners to spiritual evolution. More than that, they reveal the subtle interplay between individual choice and pre-determined conditions – 'nature' and 'nurture' – that characterizes the Jain view of the human condition. This approach overlaps closely with the concerns of modern psychologists and social scientists. The aim of 'self-conquest' is based on a holistic or *ecological* view of the self that recognizes the importance of connectedness and inter-dependence. This is a radical contrast to the narrow definitions of 'the individual' that have induced environmental and psychological crisis.

## Gunasthana Glossary: 'Scientific' Stages and 'Traditional' Jain Terms

**Fourteen Gunasthanas = Purification Stages**

Abbreviations:

WV = World-View;

EWV = Enlightened World-View;

SR = Self-Restraint, CSR = Complete Self-Restraint.

*Stage/Gunasthana:*

1. Deluded WV = *Mithyadrsti*
2. Lingering EWV = *Sasvadana*
3. Mixture of Deluded and Enlightened WV = *Mishra*
4. Non-Restrained EWV = *Avirat Samyak-drsti*
5. EWV with Partial SR = *Desa-virata*
6. EWV with Careless SR = *Pramatta-virata*
7. Carelessness-Free SR = *Apramatta-virata*
8. CSR with Unprecedented Volition = *Apurva-karana*
9. CSR with Uniformly Mild Volition = *Anivrtti-Samparaya*
10. CSR with Subtle Greed = *Suksma-moha*
11. CSR with Suppressed Greed = *Upasanta-moha*
12. CSR with Eliminated Greed = *Ksina-moha*
13. Dynamic Omniscience State = *Sayoga-kevalin*
14. Static Omniscience State = *Ayoga-kevalin*

## Chapter 9

# The Purification Prescription

### Antidotes to Karmic Influences

In the previous chapter, we described austerities as the antidote to the five karmic agents, Perverted or Distorted Views, Non-Restraint, Carelessness, Passions and Activities. In fact, the term austerities or *Tapas*, according to Umasvati, implies the development of the following qualities:

*Gupti* – Restraint
*Samiti* – Watchfulness
*Dharma* – Righteousness
*Anupreksha* – Reflections
*Samyak-charitra* – Right Conduct

Thus there are six antidotes to the five karmic agents responsible for the stoppage of karmic influx and dissociation of karmic matter. All six of these antidotes come under the heading of 'Austerities' (*Tapas*) in accordance with Axiom 4C. The reader will doubtless have noted the use of the term 'Dharma' for 'Righteousness', which includes righteous conduct by the individual and society alike. Dharma also means the natural order of the universe, understanding of which is the key to spiritual evolution. Once again, ethical and natural laws are viewed as identical in Jain science. By changing our behavior and with it our priorities and values, we bring ourselves in line with Dharma and act righteously at the same time. We also understand our true selves and the true purpose of our existence, peeling away layers of illusion and ignorance (*Avidya*).

We shall describe in more detail below the 'Six Antidotes' in relation to the fourteen purification stages. The antidotes listed (and their sub-divisions, of which more below) take us up to the sixth stage of purification. Together, they lay down a prescription for human conduct that includes restrained, healthy diet and (on occasion) fasting to purify the body, limiting speech through cultivating silence (and abstaining from gossip or hateful speech of any kind) and stilling the mind through meditation.

This chapter contains many technical details of the purification stages as envisaged and practiced within the Jain tradition. It should be read in close conjunction with the Gunasthana Glossary.

### *Astanga*: Eight Qualities of True Insight

Once the fourth stage of *Astanga* or 'True-Insight' is attained, there are eight qualities of True Insight that arise before one can rise to a higher stage on the purification axis. Four of these qualities are of a negative nature. They are:

1. *Nihsamkita:* Freedom from Doubts (regarding the Dharma, or natural order of the universe, as illuminated by Jain teachings)
2. *Nihkamsita:* Freedom from Anticipation (regarding speculation about the future)
3. *Nirvicikitsa:* Freedom from Disgust (arising from making a distinction between a pair of opposites)
4. *Amudhadrsti:* Freedom from False Notions (regarding gods, gurus and religious practices)

The other four qualities, by contrast, assume a positive form:

5. *Upaguhana:* Safeguarding (the Jain faith from public criticism, by dealing with the failings of a fellow Jain through discreet instruction)
6. *Sthitikarana:* Promoting Stability (by making others more

certain of their faith when they are tending towards scepticism and doubt)

7. *Prabhavana:* Illumination (by positive actions which promote Jain philosophy and science)
8. *Vatsalya:* Disinterested Love (involving a selfless devotion to the ideal of Moksha and thus great devotion to the monks – or *Muni* – pursuing this goal)

The eight qualities of Astanga are expressed in terms both of religious devotion (Jain practice) and the quest for knowledge (Jain science). They also illustrate an aspect of Jainism that can seem paradoxical, even contradictory to outsiders. This paradox is found in the ability to combine inner certainty with continuous questioning, stability with continuous development and change. For in the Jain system of logic, multiple viewpoints are recognized as aspects of a larger truth. When we realize this, we learn to approach philosophical – and religious – questions in a radically different way. From the Jain standpoint, certainty and clarity emerge from the absence, the shedding, of 'one-sided' dogmas.

## The Fifth Stage for Jain Lay People

Purification Stage 5 involves the following eleven model steps of renunciation for lay people. The most important stage is the taking of the Lower Vows prescribed for a lay man or woman. Of these, the five Lower Vows are the most important. These are:

1. Avoiding injuring beings having two or more senses: *Ahimsa*
2. Being truthful: *Satya*
3. Refraining from stealing: *Asteya*
4. Avoiding promiscuity and sexual activities outside a committed relationship: *Brahmacharya* (also sometimes called *Brahmavrata: Anu Brahmavrata* for the Lesser Vows,

*Maha Brahmavrata* for the Greater Vows)
5. Limiting one's possessions: *Aparigraha*

This last of the eleven sub-stages culminates in preparation for the next stage, that of the Muni or monk.

In Jain practice, the Lesser Vows adopted by lay men and women are known as the *Anuvratas*. In essence, they are modified versions of the practices followed by male and female ascetics, which are known as the Greater Vows or *Mahavratas*. The Anuvratas are rules of conduct adapted to living and participating fully in the community. They point towards a way of life that could be referred to in secular parlance as 'ecological' or 'sustainable', working with the grain of nature rather than seeking to 'conquer' or rise above it. In other words, the Anuvratas are intended to encourage the individual and the wider Jain community to live in recognition of the Dharma – and hopefully inspire others to do so by quiet example. These vows could be said to fulfill the original meaning of economics, which is derived from the ancient Greek words for 'law of the household': good housekeeping and a restrained but fulfilling way of life. The Mahavratas, by contrast, point the way to an ascetic existence based on withdrawal from the responsibilities of household and community in favor of ultimately higher goals of contemplation, study and spiritual development.

The two gradations of vow are also mutually dependent. Without the householders who follow the Lesser Vows, the framework for practice of the Greater Vows would not exist. Or, to put it another way, the economic and cultural support network for the ascetics could not be sustained. At the same time, the Mahavratas and those who follow them point householders towards the idea that there are goals beyond material success and even familial happiness. The wealthiest and most successful Jain businesspeople and professionals are reminded of the ideals of simplicity and renunciation, which bring them closer to

Moksha. This is why, although their success and stability is valued highly by the Jain *sangha* (community of adherents), such men and women hope that they will have the chance to practice the *Mahavratas* in a future existence. They therefore donate generously to the ascetics and to educational or charitable foundations, with the concept of charity extending to the environment and animal welfare as well as human-centered causes.

## Stage Six and Jain Ascetics

Stage 6 involves following the Higher Vows which involve tougher austerities. These are extensions and additions to the Lower Vows 1-5 above and, in particular, include renunciation of one's possessions and an end to conventional domestic life.

The overall aim is to minimize the extent and the frequency of activities which would lead to additional karmic matter being taken in through the arousal of new passions. The practices of an ascetic man or woman prepare an aspirant for the advanced stages of meditation and understanding through which the influence of karmic matter is dispelled. Once karmic matter has been expelled from the Jiva, liberation or Moksha is achieved. In Jainism, unusually for an Indic tradition, there are more ascetic women than men and this has been the case since the time of Mahavira, whose openness to female as well as male aspirants was greater than his contemporaries, including Gautama Buddha.

The practices followed by an ascetic can be summarized as follows:

1. *Restraint – Gupti:* There are three restraints (known popularly as the 'Three Gupti', which involve a progressive curbing of the activities of the body, mind and speech, i.e. aiming for single-mindedness. The three Gupti are therefore *Mana Gupti* (Regulation of the Mind), *Vachana Gupti* (Regulation of Speech) and *Kaya Gupti* (Regulation of

Bodily Activity).

2. *Watchfulness – Samiti:* There are five Samiti, or forms of watchfulness, involving positive caution in one's activities:
   A. *Irya Samiti* (or *Iriya Samiti; Iryasamiti*): taking care when walking to avoid killing or hurting small creatures
   B. *Bhasa Samiti:* attempting to speak truthfully and as little as possible, avoiding the eight faults of speech: pride, deceit, greed, laughter, fear, gossip and slander
   C. *Esnna Samiti:* accepting alms in such a way that there is no feeling of self-gratification
   D. *Adana Nikshepana Samiti:* care in picking up and putting down objects so that no form of life is disturbed or crushed
   E. *Utsarga Samiti:* care in performing the excretory functions so as not to disturb living things.

The first and best known of these, Irya Samiti, encompasses all five precepts of 'Watchfulness'. It means more than merely 'care in walking' but extends to Careful Action in general. Careful Action is the principle underlying all aspects of the Jain way of life, because every action can have unforeseen consequences, sometimes generations from now, or unintended effects on other forms of life. Moreover, all action is by definition 'karmic'. It follows that lay Jains who have adopted the Lesser Vows also practice a modified form of Irya Samiti in their daily living. Ascetics take the principle as far as they can to its logical conclusion, hence the popular images of monks who cover their mouths with cloth and sweep the ground before them with small brushes to avoid injury to minuscule (but complex) life.

3. *Dharma – Righteousness:* One cultivates ten rules of righteousness to reinforce these practices. These are perfection in forbearance, modesty, uprightness, truthfulness, purity, restraint, austerity (related to intense meditation),

renunciation, detachment and continence. The rules of righteousness are known as the *Dasa-Dharma*.

4. *Anupreksha – Reflections:* The twelve mental reflections engaged upon are given below.

The traditional approach makes them appear somewhat 'negative' in character. However, Gurudev Shree Chitrabanu, founder of the Jain Meditation International Center in New York City, has reinterpreted them in a more positive manner, in keeping with the interests and demands of modern urban life. The twelve *Anupreksha* may therefore be presented as follows, paraphrasing and in places adapting Chitrabanu's 1981 book, *Twelve Facets of Reality: The Jain Path to Freedom*. The relevant English phrase is stated and summarized, with the original Sanskrit concept following in italics.

1. *Impermanence.* There is impermanence of everything surrounding one but there is unchanging soul beneath the changing body: *Anitya*
2. *Helplessness.* We are helpless in the face of death but the inner invisible force always lives: *Asarana*
3. *Cycle of Rebirth.* Liberation from the cycle of rebirth is possible: *Samsara/Sansar*
4. *Aloneness.* There is the absolute solitude of each individual as he goes through this cycle and therefore one should achieve dependence only on oneself: *Ekatva*
5. *Beyond Body.* The soul and body are separate and we are more than merely corporeal. We must seek the true meaning of life through the existence of the soul: *Anyatva*
6. *Impurity.* Even the most physically attractive body contains impurity: *Asucya*
7. *Karmic Fusion.* Karmic influx happens and how to stand apart and watch the inflow: *Asrava*
8. *Karmic Shield.* Influx may be stopped and how to close the

window when the storm, in the form of the Four Passions, is about to come: *Samvara*

9. *Total Karmic Decay.* Karmic matter within the soul may be shed so that the soul may be cleaned to move towards permanent reality: *Nirjara*
10. *Universe.* The cosmos is eternal and uncreated (by an external agent), hence each person is responsible for his own salvation – for there is no God to intervene: *Loka-akasha*
11. *Rarity of True-Insight.* True-Insight is rarely attained and human embodiment bestows the rare privilege and opportunity to attain Moksha: *Bodhi-durlabha*
12. *Truth of Jain Path.* The truth of the teachings of the Tirthankaras which lead to the goal of eternal peace through understanding one's own true nature: *Dharma*

Most of these concepts have already been explored. Yet a few of the Anupreksha are open to misinterpretation. Precept (4), for example, is not a cosmic endorsement of Margaret Thatcher's notorious claim that 'There is no such thing as "society", only individuals and their families'. The individual is 'alone' only in so far as he or she is not directly answerable to a supreme being. Accordingly, the individual's spiritual development is ultimately his or her own responsibility. However, despite the apparently uncompromising language, this concept of individual responsibility does not absolve the individual from commitments to others. On the contrary, self-realization also means realizing that there is a far wider concept of 'society' than that which is generally conceived of in modern thought, for it includes all beings and ecosystems.

Equally, Precept (12) need not exclude followers of other spiritual paths, or indeed atheists and humanists. The pluralist logic of Jainism acknowledges that understanding of Dharma can arise from other cultural traditions. The 'truth' can be approached from many different angles and so Jains do not have a monopoly of truth. It is possible to be aware of 'the truth of the teachings of the Tirthankaras' without even being consciously aware of them.

5. *Parisahajaya – Mastery over Afflictions:* The 'Mastery over Afflictions' consists of over twenty-two typical hardships which should be meditated upon; examples are hunger, thirst, cold, heat, insect bites and ridicule. It is aimed primarily at Muni living in the Indian environment and climate, but it still has relevance to those practicing austerities elsewhere.

It should be noted that the Gupti, Samiti and Dharma precepts are intended guidelines for lay people to work towards according to their abilities and circumstances. There is no external authority, divine or human, that demands 'perfection' and judges 'imperfections'. Lay people have flexibility in their practice, but use the precepts of spiritual practice as a form of 'gold standard' to measure their actions and attitudes against. There are, however, community-binding practices such as fasting on special days that lay men and women are expected to observe. Monks, by contrast, are expected to follow these guidelines to near-perfection: the pressure to do so can come from the lay community, which sustains them economically, but it arises primarily from within.

## Meditation

To move on to higher stages, one uses advanced meditation comprised of *Dharmadhyana* or 'Virtuous Meditation' and *Sukladhyana* or 'Pure Trances'. Virtuous Meditation entails deep contemplation[1] for up to forty-eight minutes on the following themes:

The Jain teachings on the Nine Reals or *existents* (Tattva)
The means by which to assist others
Karmic decay/emission
The structure of the universe

During such periods, Carelessness is suppressed and the contemplative temporarily attains the seventh purification stage. As he or she enters and leaves the contemplation periods, there is alternation between the sixth and seventh stages.

These contemplations, free from Carelessness, are considered to be preparatory to, but do not themselves lead to, the defeat of the subtle passions. Only with the attainment of the eighth stage, 'Unprecedented Spiritual Progress', can one be sure of reaching the highest step leading in the end to Moksha. This can occur only through Pure Trances, of which there are *four* types:

1. Pure Concentration on nature and multi-modal aspects of the six existents
2. Pure Concentration on a unimodal aspect (i.e. a single facet) of an existent (Tattva)
3. Transcendental state of subtle movement
4. Transcendental state of absolute immobility

The first two Pure Trances operate in the eighth, ninth and tenth purification stages, during which both the subsidiary passions and very subtle passions are progressively suppressed or eliminated. Eventually the Jiva will gain sufficient energy to mount the ladder and so eliminate rather than suppress passion at each stage. Thus the eleventh stage will be omitted and the soul will enter the twelfth stage. The purification of the soul is now at its highest which instantaneously moves to the thirteenth stage: Dynamic Omniscience State.

A few moments before physical death, the last two Pure Trances are employed in succession and this sets the irreversible process of reaching the fourteenth stage. Through the third Pure Trance, one completely stops the activities of the body, mind and speech except for the regulatory processes of breathing and heart-beat. Through the fourth Pure Trance, even the regulatory processes are stopped, and the soul attains Moksha.

We have concentrated on two positive meditations. There are also negative meditations or negative mental states. One is *Artadhyana* or *Mournful Meditation* which is brooding on something disagreeable, e.g. the loss of loved ones or the loss of valued possessions. This mental state is a state of anxiety or worry that impedes greater awareness and self-knowledge. The other is *Rutradhyana* or *Wrathful Meditation* which is dwelling on the perpetration of violence, falsehood, theft, sexual improprieties, as well as the extreme preservation of one's possessions. These states should be overcome if we are to move beyond the sixth stage of purification.

## *Tri-Ratna*: The Three Jewels of Jainism

The Axioms of Jain science may be summarized into the following single scriptural verse from Umasvati:

> 'Right Faith, Right Knowledge and Right Conduct constitute the path to Moksha'.

Right Faith, Right Knowledge and Right Conduct are called the Three Jewels of the Jain path, or *Tri-Ratna*. The Sanskrit terms for them are *Samyak-darshana, Samyak-jnana* and *Samyak-charitra* respectively. They are attained sequentially: Right Faith is the first to be achieved: this occurs at purification stage 4. Right Conduct is achieved at stage 8 and Right Knowledge at stage 13. The Three Jewels are usually depicted with the svastika beneath them to indicate the four 'states of existence' or mental states. In the words of Umasvati, author of the *Tattvartha Sutra:*

*Right Faith (= True-Insight) consists of belief in the soul, karmic matter and the other seven Reals, the Right Knowledge is their comprehension and Right Conduct is austerity.*

Right Knowledge also emphasizes Anekantavada, or non-absolutism, achieved through the intellectual method of Syadvada and a mental attitude that is open to 'multiple viewpoints' as a means of arriving at a more 'rounded' and multi-dimensional truth. It is said that the order of mental and emotional development is 'first knowledge, then compassion'. The purpose of knowledge, indeed, is to increase compassionate understanding or *empathy*, as it is now widely known. Without compassion, knowledge is sterile and only partially complete. Without knowledge, compassion becomes mere sentimentality.

Right Conduct is austerity or Tapas, already described in some detail, but austerities practiced without understanding of the reasons behind them have little effect: 'If a man without Right Knowledge were to live on only a blade of grass once every month, he would not be entitled to even a sixteenth part of merit'.

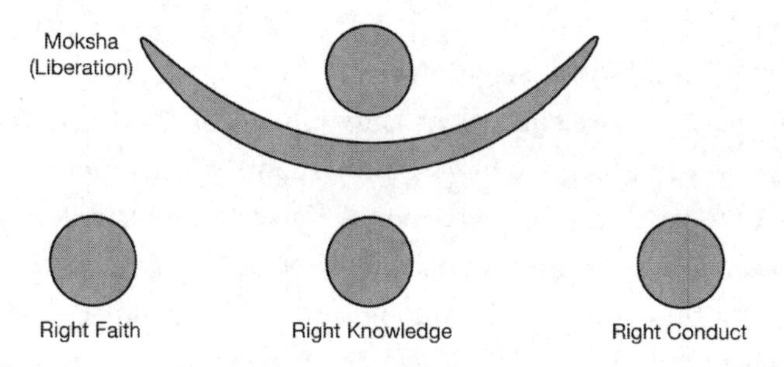


Image 4: *Moksha and the Three Jewels*

Within the Jain tradition, spiritual progress is often likened to the making of the clarified butter known as ghee:

| Ghee Stage | Parallel | Purification stage number |
|---|---|---|
| 1. Collection of milk | Right Faith* | 4 |
| 2. Heating of milk | Austerity rituals | 5 |
| 3. Cooling of milk | 'Cool' mind (meditation) | 6 |
| 4. Addition of culture | Right Knowledge | 6 |
| 5. Stilling of mixture (6 hours) | Right Conduct | 6 |
| 6. Churning to produce butter | Advanced meditation | 7-11 |
| 7. Placing of mixture on fire | Fire = Advanced Meditative Trances | |
| to produce Ghee | Ghee = Pure soul (Jiva) | 12 |

*Right Faith, in this context, means awareness of the pure soul (Jiva) as the state of liberation to work towards as far as possible.

## Spiritual Progress and Driving: A Useful Analogy

The spiritual progress of the individual through the fourteen stages (Gunasthanas) can be illustrated quite effectively by the process of learning to drive a car and then becoming increasingly skilled. The British driving test standard (which we shall take here as our model) does not require complete perfection. It is assumed that further improvement will take place after the test is passed. However, from the Jain standpoint the individual is a 'learner' until Moksha is attained.

Stages 1 to 4 are equivalent to gaining a correct understanding of the use of the car, viewing it in this context as a useful machine which must be driven in such a way that one avoids risk or injury to oneself or others. A person who believes, and can put into practice, this concept should be able to pass the British test. However, subsequent improvement may take place in order to become an advanced motorist, rather like the path of the monk.

Stages 5 and 6 involve the attainment of full restraint: although in control of the car, one must nevertheless avoid accelerating quickly, braking hard, or flashing lights / hooting unnecessarily. The phenomenon popularly known as 'road rage' is an especially virulent expression of the Kashaya! The Three

Restraints (Gupti) mean a progressive curbing of the activities of the body, mind and speech so that one acts instinctively, without conscious thought. Stage 7 is the achievement of Watchfulness, i.e. using mirrors, indicators, lights, etc. when necessary so that no other user is made anxious by one's bad driving, even though an accident would have been unlikely. Also, one is alert at every moment so that enough time is allowed to take corrective action on the bad driving of others, etc.

Stages 8 to 12 involve the reduction and elimination of passions in driving. These are the most difficult faults to remove and involve feelings such as impatience in a long traffic jam and uneasiness when repeatedly overtaken, even though driving at just below the maximum legal speed. These passions may smoulder and only arise occasionally, since they are usually held in check. At stage 13, one has reached the position of causing the minimum possible danger on the road. Stage 14 is the beginning of the cessation of activity (yoga): this means that one sees that doing without a car altogether eliminates this contribution to danger. It is worth noting that one of these activities takes place whilst the car is stationary with the engine not running, i.e. there is no yoga. Therefore this is not a perfect analogy, but an approximation of the purification process.

We can also illustrate the use of the five Samiti (forms of Watchfulness) through this analogy. The first Samiti is rather like driving so as not to hit birds, rabbits or other creatures on the road. The second is similar to reducing conversation in the car to decrease distractions. The third can be likened to not drinking and driving, thereby retaining full concentration. The fourth is akin to looking around the car before starting and choosing a parking space carefully so that no child or animal is hit. The last Samiti resembles the avoidance of running the car engine in a confined space where people may be affected by the exhaust fumes.

Needless to say, the Jain ethos leaves little room for materialistic interest in cars as luxury items or status symbols – although in practice by no means all Jain householders adhere to this aspect of Aparigraha (non-possessiveness)! Driving is also, in strict Jain terms, a yoga which should be avoided as much as possible, especially in the light of environmental concerns. To return to our analogy, a deepening spiritual commitment could be likened to cutting out unnecessary journeys and 'downshifting' from a luxury brand to an electric vehicle. Nonetheless, driving is a useful analogy with spiritual development because it is a logical, indeed scientific process. Science is, after all, a broad concept, embracing knowledge in all its forms and facets.

## Epilogue

# Why Jain Science?

*Science without religion is lame*
*Religion without science is blind*[1]
Albert Einstein

In the chapters above we have explored the relationship between the Jain spiritual path, with its accompanying beliefs and practices, and the process of scientific inquiry, with its quest to expand the frontiers of objective truth. We have sought to demonstrate that the Jain approach to spiritual development is founded on a rational view of the universe and its workings. At the same time, the scientific quest is incomplete without a spiritual component, that is to say awareness that there are 'more things in Heaven and Earth' than the exclusively material or tangible. However, our intentions go further than the 'rationalization' of spirituality or the 're-enchantment' of science. For we aim to present Jainism as a unified spiritual system or Holistic Science, encompassing the physical and spiritual realms alike and abolishing the distinction between the two. In this context, Jainism – or more accurately, the Jain Dharma – is more than a religion in the discrete and circumscribed sense in which we have come to know the term in the West. It is a system of thought, a way of life and a way of looking at the universe. Beyond that, it is the rich, subtle and still little-known culture that we define as *Jainness*.

This summary points us towards the question of *why*. Why should the Jain perspective be of interest to readers when other Indic traditions, with many more adherents, are already being explored in depth, as any casual visit to a western bookstore will indicate? Moreover, through works such as *Blackfoot Physics* (Peat, 2005) and *The Tao of Physics* (Capra, 1975), the idea of a connection between cutting-edge science and eastern or indigenous spiritual pathways is readily understood. Through the science of ecology, we become aware, both intellectually and materially, of the interconnectedness of living systems, however intricate or improbable those links might be. We know that even the tiniest particles play a critical role in sustaining life in the universe and even the smallest organisms are essential for the preservation of life.

Equally, we are aware that such small organisms or particles are as complex in their structure and are in their own way as 'intelligent' as human beings, even if they lack the peculiar human capacities for creativity and destruction, heightened self-awareness and extreme moral blindness.

We are even becoming familiar with the idea that the universe, long conceived of by rationalists as a monolithic structure, can consist of multiple parts that interact and overlap. 'Multiverses' and 'parallel universes' have become part of popular culture (much as the different levels of Heaven and Hell once were, and remain so for some cultures, such as the Tibetan Buddhists). At the point where theoretical physics and the humanities intersect, we find the 'implicate universe' of David Bohm, a physicist who engaged in dialog with Native American elders and the Hindu mystic Krishnamurti. For Bohm, the universe in its complexities, surprises and wonders is best conceived as an 'implicate' or enfolded order, in which 'within each object can be found the whole and, in turn, this whole exists within each of its parts'.[2]

This view of the universe is wholly consistent with the Jain conception and is arrived at through the same combination of rational and intuitive thought. It is, in short, a 'many-sided' view of the universe. And Many-Sidedness is not in itself a new idea either, for the twenty-first-century West. As early as 1872, the American poet John Godfrey Saxe (1816-1887) transcribed a version of the story of the blind men and the elephant (see Chapter 2), describing it as 'A Hindoo (sic) Fable'[3]:

## The Blind Men and the Elephant

*It was six men of Indostan*
*To learning much inclined,*
*Who went to see the Elephant*
*(Though all of them were blind),*
*That each by observation*
*Might satisfy his mind.*

*The First approached the Elephant,*
*And happening to fall*
*Against his broad and sturdy side,*
*At once began to bawl:*
*'God bless me! – but the Elephant*
*Is very like a wall!'*

*The Second, feeling of the tusk,*
*Cried: 'Ho! – what have we here*
*So very round and smooth and sharp?*
*To me 'tis mighty clear*
*This wonder of an Elephant*
*Is very like a spear!'*

*The Third approached the animal,*
*And happening to take*
*The squirming trunk within his hands,*
*Thus boldly up and spake:*
*'I see,' quoth he, 'the Elephant*
*Is very like a snake!'*

*The Fourth reached out his eager hand,*
*And felt about the knee.*
*'What most this wondrous beast is like*
*Is mighty plain,' quoth he;*

*'Tis clear enough the Elephant*
*Is very like a tree!'*

*The Fifth, who chanced to touch the ear,*
*Said: 'E'en the blindest man*
*Can tell what this resembles most;*
*Deny the fact who can,*
*This marvel of an Elephant*
*Is very like a fan!'*

*The Sixth no sooner had begun*
*About the beast to grope,*
*Than, seizing on the swinging tail*
*That fell within his scope,*
*'I see,' quoth he, 'the Elephant*
*Is very like a rope!'*

That this exposition was published to popular acclaim in the United States some 140 years ago (hence the archaic language) is evidence that the ideas associated with Jain Science are already contained within the cultural landscape.

The awakening of green spirituality is complemented in scientific theory by ideas such as the Gaia hypothesis, which conceives of the Earth as an integrated, ultimately self-regulating whole and reminds humans that they are relative newcomers and so far only bit players in the history of life on Earth.[4]

The philosophical movement known as Deep Ecology transcends campaigning environmentalism because it asks more fundamental questions about the relationship between human beings and the natural world and reveals the psychological scars that arise from the division between them.[5] These scars are reflected in the damage inflicted on the environment in general, and other species in particular, by humans afflicted with delusions of superiority. Understanding of the threat posed to all life (including our own) by the processes of continuous economic expansion has coupled with a moral and aesthetic sense that consumption for its own sake is ultimately sterile. The result is a growing reassessment of what it is to be human. This is pointing towards an inclusive humanism that takes account of humanity's dependence on other living systems, rather than 'dominion' or mastery over them. Such a radical change or paradigm shift requires a more inclusive view of the human individual, whereby the individual is seen (like every other life form) as unique and precious, but deriving full individuality through co-operation rather than competition with others. Learning to co-operate with each other is part of the process of learning how to co-operate with the rest of nature.

This is where Jain thinking comes in. The Dharma, or order, of the Jains clearly combines reason and ethics. It reminds us that when a species is lost to extinction, we lose a part of ourselves. Equally, when we cut down an area of forest or poison a river, we are choking off the possibilities for ourselves. From the Jain standpoint, anything that contains Jiva, or life energy, has the possibility of spiritual evolution, which gives it kinship to humanity.

The idea of biological evolution from one life form to another has never seemed strange to Jains, because it is merely a reflection of spiritual development. The lesson that derives from both forms of evolution, however, is not human supremacism but Careful Action, the conscious choice to avoid using our power or intelligence to inflict harm on fellow beings. This principle arises from a sense of the limits as much as the extent of human faculties. An ethical standpoint of avoiding harm unites with a viewpoint based on rational self-interest. The infliction of harm is materially damaging and life-threatening to ourselves, especially when it involves damage to the conditions on which we depend for survival. At the same time, destructive patterns of behavior and thought do psychological and spiritual damage to us as individuals and as a species. Learning to live within limits, through carefully evaluating our actions (and the thoughts that inspire them) is viewed as the most powerful indicator of human progress.

Ahimsa, the Jain principle of non-violence, does not mean simple non-involvement for lay men and women, but active compassion directed at all forms of life. It challenges the attitude of fatalism and pessimism that accepts, for example, that 'all bear species will become extinct this century' or that '(human) inequality is inevitable'. At an even deeper level, Jain Dharma issues a challenge to the ideologies of conflict, economic expansion and human supremacism that still remain dominant, in the political sphere most of all. For the modern West (and those cultures under its spell or sway), it offers an alternative method of forming thoughts and ideas. The many-sided approach is about 'both/and' rather than 'either/or', convergence and synthesis rather than 'winning arguments' and an inclusive vision of the truth rather than its reduction to narrow certainties. It recognizes that knowledge and truth are multi-dimensional, in the same way that the universe can be seen as having many layers or 'folds'.

More than that, the concept of Many-Sidedness reminds us that each living being has its own limited but valid version of the truth that contributes to the sum total of knowledge of the universe. Each of us, in our own way, is engaged in a quest for knowledge. We begin to remove the blinkers from our eyes when we understand how little we know, however apparently advanced or erudite we might be. Equally, we need to see that every organism, however seemingly elementary, has its own *Naya* or viewpoint that is as valid as ours and from which we could learn.

This Jain view of the pursuit of knowledge – science in its original sense – is very different from that to which we have become broadly accustomed. Yet we are collectively beginning to feel the limitations of the adversarial mentality. Increasingly, we are also confronting its consequences in environmental despoliation, species loss, the destruction of whole human cultures, failed political ideologies and entirely unnecessary economic deprivation. The Many-Sided approach provides an alternative strategy for formulating and applying ideas. Rather than expecting us to 'become Jains', it allows us to find repositories of wisdom within our own civilization and just as importantly within ourselves. It frees us from the constraints of either/or logic and gives us more efficient tools for living in a world where interdependence is the only practical way to survive. The principle of Anekant, or Non-Violence of the Mind, is the basis for Jain Science. It is also an under-used intellectual resource. Given a wider application, it might enable the state of fermentation in the physical sciences to extend itself to social and political thought.

# Appendix 1

## A Brief Life of Mahavira

Mahavira was born in 599 BCE in Kundagrama, then a large city in Northern India, near the modern city of Patna. His father was King Siddhartha and his mother Queen Trishala. His original name, Vardhamana, which means 'ever growing', was given to him because everything in the kingdom became abundant during the period of his mother's pregnancy.

Vardhamana soon developed a great sense of understanding and rapport with animals. Even in his childhood he courageously (and non-violently) subdued a terrifying snake. He also calmed an elephant that had gone on the rampage, and stopped it from doing further damage. A fight with a large bully led to the name Mahavira or 'Great Hero'.

The young Vardhamana almost certainly received the typical training of a prince of that period, in disciplines ranging from literature, political science (or statecraft) and mathematics to archery. He was highly intelligent and at an early age his teacher confessed that Mahavira was ahead of him in knowledge.

According to the Svetambara Jain scriptures, Vardhamana led a conventional domestic life and married Yasoda. The couple had one daughter, Priyadarshana. The Digambara Jains, however, believe that he was never married and had no children. According to one popular version of the Tirthankara's life, the twenty-eight-year-old prince left his palace one day and saw a slave being whipped by his owner. This opened his eyes to the social injustice of his contemporary society, where the Dharma or natural law was routinely ignored and violated by exploitation of the poor by the rich and the persistence of illiteracy and ignorance. In this way, the transformation began from Vardhamana, the young prince sustained by material wealth, to

Mahavira, the spiritual conqueror who cultivated inner riches.

Mahavira saw that the root of social injustice (including the exploitation of animals and nature) was the loss of spiritual wisdom: a failure to understand the true nature of the universe (Dharma) and the true nature of the self (Jiva). This was a collective failure, but its roots lay in the inner life of each individual human being. Such realizations led Mahavira to wish to pursue a spiritual path that involved renouncing material ties, which were illusions, and as part of this leaving his privileged surroundings. He did, however, have a deep feeling of consideration for his parents and this moved him to vow that he would not renounce domestic life until they had both passed away.

After the death of his parents, Mahavira waited until some two years later when their loss had become bearable to his elder brother, and then asked his permission to leave the palace. It is believed that during these last two years in the palace he spent a considerable time in self-analysis, rather than in mundane, everyday pursuits. It should be noted that the Digambaras believe that Mahavira took ascetic vows and became a monk while his parents were still alive. He then left home to search for the root of all problems: to understand human nature and to study the universe in general.

Mahavira devoted the subsequent twelve and a half years to his research with intense single-mindedness. As he felt that it would assist his meditation, he lived very frugally, wandering from place to place wearing only a single garment and frequently fasting. He also reduced his other needs, for example by removing his hair by hand. So intense was his concentration on his goal, that when his garment was accidentally caught on a thorn bush and pulled off, during 13 months of renunciation, he remained naked. (However, according to the Digambaras, he removed his clothes at the time of renunciation.)

Another incident showing his firmness of purpose relates to how he was meditating in a standing posture on a farm and the farmer, who had his cows grazing around him, asked him to look after them whilst he was away. Since he was in a state of deep meditation, Mahavira did not notice that the cows were wandering away. When the farmer came back, he asked Mahavira about the missing cows and, since he was under a vow of silence, he did not reply. The already upset farmer was further infuriated and he hammered two wooden nails through Mahavira's ear to punish him for this lapse. But even that action did not break Mahavira's silence, and Mahavira remained compassionate towards him.

It is said that he remained in total solitude until Mikkhali Goshala, who had heard of Mahavira's outstanding abilities, searched and found him. Goshala was a travelling story-teller and a follower of the fatalist doctrine of the Ajivika sect of which he later became chief spokesman. It is related that they were together for six years during which time Goshala became thoroughly acquainted with Mahavira and his abilities. Mahavira described the six months of austerities, which he thought necessary for the attainment of these abilities.

Mahavira finally attained *kevalanana* (Dynamic Omniscience State) precisely twelve years, six months and fifteen days after starting out on his search. Thus he was able to comprehend the mechanism of the universe as a whole and human nature in particular, which led him to the root of all problems.

Having left his princely state in pursuit of his goal, on his enlightenment Mahavira came back to share his knowledge with the community. The event of coming back is far more significant than his search. He gave his first sermon to an audience that included Indrabhuti Gautama, who was well versed in Hindu scriptures and extremely proud of his knowledge. Through this encounter Gautama became his chief disciple (*ganadhara*). Eventually he had eleven ganadharas as his inner circle. He had a great natural organizational ability and as his followers grew in number, he formed *'tirtha'* (the order) of monks, nuns, lay men and lay women. This can also be referred to as the Jain *Sangha*, or community. Also his daughter, Priyadarshana, who was married to Jamali, eventually became a follower of Mahavira.

To distinguish his ideas clearly from the prevalent influence of Hinduism, Mahavira developed a very versatile talent for coining new terminology, for example the lay followers were called *shravaka* (male) and *shravika* (female), those who are attentive (right) listeners, and monks were called *Shramana*, that is, laborers on the spiritual path. He vigorously reaffirmed the concept of autonomous self-responsibility, removing the idea of a God who influences the day-to-day activities of everyone. Further, he claimed 'Every man has a right to and could attain nirvana by his own effort without the help of any supreme authority or mediatory priest.' This idea of self-responsibility should not be confused with the modern cult of 'personal autonomy', where the self is defined narrowly and in almost exclusively materialistic terms. For Mahavira and his Jain tirtha, self-realization involved understanding that all 'selves' share the same properties and are interdependent.

Mahavira preached the equal value of all living beings, including the equality of all humankind and hence the abandonment of slavery, the caste system and animal sacrifices. These violations of Dharma distort human relationships and the balance of nature. The leader of Mahavira's order of nuns, Candana, came from a slave background, slavery having become prevalent in the North India of Mahavira's day. At the other extreme, one of the kings of that time, Bimbisara, became a staunch follower, perceiving the value of Mahavira's teachings.

One of the revolutionary contributions of Mahavira to Indic spirituality was to change the Hindu recommendation that monkhood should not commence before the latter part of one's life. He introduced the idea that there need be no specific time ldly renunciation, with gradual transformation l for those who are not ready for total renunciation at an early age.

Another outstanding feature of Mahavira's character was that he was the perfect living image of compassion towards all forms of life. An example quoted is of a cobra called Chandkosia, who had been withstanding all who had tried to cross its path. One day, the cobra bit Mahavira, but such was his knowledge that he could see, through the cobra's past lives, how it had developed such a nature and he had great compassion for it. Such was his compassion that it was as though milk flowed through the wound and the injury became secondary to the concern that Mahavira had for the cobra's well-being.

Ultimately, Goshala turned against Mahavira and tried to intimidate him with a curse, saying that he would die of a fever within six months. Mahavira did become ill but eventually recovered. The death of Goshala shortly afterwards gave the impression that the curse had returned to its source. However, Mahavira himself was always against magical or Yogic power for self-interested ends.

Mahavira continued to teach and practice the three jewels up to the time of his physical death. Various fundamental teachings and practices, with only minor variations, are still prevalent among Jains. In particular, all Jains celebrate the festival of lights (*Diwali*) because on the day of Diwali Mahavira achieved Moksha, whereas on the same day his main disciple Gautama became a *kevalin*: one who has attained the state of omniscience.

## Appendix 2

# Comparative Views

*Indian culture has been enriched by extraordinary spiritual leaders, thinkers and saints. ...In mathematics, astronomy and chemistry, the chief ingredients of ancient science, Indians discovered and formulated much and anticipated, by reasoning or experiment, some of the scientific ideas and discoveries which Europe arrived at much later.*

Acharya Mahapragya (10th Acharya of the Svetambara Terepanth school) and A.P.J. Abdul Kalam (President of India, 2002-2007)[1]

The Jain tradition is less well known in the West than (for example) its Buddhist counterpart. However, at this juncture we believe that it is especially well-placed to guide us through some of the ethical dilemmas we face, both as individuals and as a human species that has assumed such extreme power to create and destroy.

The economic and technological hegemony of the West, seemingly assured for so long, is challenged increasingly from the East, not least by a resurgent India. Against this background, the 'mainstreaming' of Jain science could foster a new spirit of understanding and equal exchange. This would contrast positively to both the cultural imperialism of the recent past and the destructive rivalries of the present.

The Jain path cannot be slotted neatly into the recent awakening of green thought in the West, especially, and the campaigns that have stemmed from it. From a Jain perspective, for example, the starting point for environmental consciousness need not be the effect of human-impacted climate change on human communities, still less human-made economies. A more logical beginning, for many Jains, is the human expansionist tendency and the consequent terrorism and destruction inflicted on other species. When this is addressed, and the attitude of human arrogance or false consciousness gives way to an attitude of restraint and care, only then can the problems that specifically impact on humans be approached effectively.

George Bernard Shaw, the Irish thinker and playwright, (see epigraph to Introduction) was at the forefront of the progressive and rationalist currents of the day, based on the assumption that humanity is the measure of all things.

As a Fabian socialist, he believed in achieving social justice and economic democracy through stages of centrally directed reform based on rational planning. In 1895, he was one of the founders of the London School of Economics and Political Science, which in its early days reflected this highly optimistic view of humanity.

At the same time, Shaw was aware that along with its many positive qualities and profound benefits of this world view (unrivalled by successor doctrines of political economy), it was in certain respects limited and one-sided. He became increasingly conscious that progress was far from a simple narrative, a line moving forever forwards, but more as the Jains and other Indic schools see it, as a continuous spiral motion, or a series of forward and reverse cycles. His critique of human arrogance, expressed in plays such as *Arms and the Man* (1894) and *Man and Superman* (1903), enabled him to make connections between colonial subjugation (the conquest of one human group by another, based on notions of racial or cultural superiority) and the exploitation of other species (based on notions of human superiority and dominance). In modern parlance, he connected racism and species-ism. Jains have always made similar connections and this is why their system of thought appealed to the prescient Shaw. Today, the problems of human-centered (in other words one-sided) notions of progress are more urgently apparent. A new generation of Shaws is hungry for ideas and for practical ways to reintegrate humanity with nature. We hope that this book will at least make a modest contribution to this task.

The starting points for what we might call Jain humanism are first that humans have an (as yet) unparalleled capacity for spiritual wisdom, but secondly that the human condition contains inherent dangers. Moreover, these dangers are of such magnitude that we must be continuously aware of their existence and consciously avoid them. In this crucial respect, the Jain version of humanism differs from that of the western tradition, guided by classical antiquity but molded by the European Renaissance and Enlightenment. Western humanism's great strengths have been its emphasis on scientific reason and artistic creativity, its conception of individual liberty and human rights, and the promise of emancipation from bigotry and superstition. Yet its great parallel weaknesses have been the assumption of human supremacy over the natural world, a mechanistic and unfeeling form of rationalism, doctrinaire forms of logic that exclude other possibilities and a denial of the spiritual element in all beings.

The promise of reason thereby gives way to a barren and materialistic conception of the universe, while the belief in human creativity is used to create an artificial division between nature and humankind, so that we lose our sense of continuity with the environment.

These aspects of western thought are proving to be untenable, in politics and economics as much as in science. Aside from the increasingly well-documented effects of human activity on the planet – activity based on dominating rather than co-operating with the natural world – we are finding that the sense of disconnectedness from the environment is having severe psychological consequences for individuals. Thus the ecological crisis is also a psychic or spiritual crisis. In this context, the Jain approach to human potential would act as a healthy corrective to the current materialistic paradigm, which has solidified largely through lack of coherent challenge for the past two and a half centuries.

Western rationalism tends towards linear thinking: the measuring of progress, whether social or evolutionary, in terms of a straight line moving inexorably forwards. It stresses the discrete qualities that separate and compartmentalize, isolating one entity from others. Non-western paradigms, by contrast, usually stress the importance of cycles and focus on continuities and common properties. They are more interested in shared characteristics than differences, whether these characteristics are shared between different aspects of observable nature or the material and spiritual worlds, which are viewed as continuous. The non-western viewpoint – as we have called it for convenience – is most pronounced among 'indigenous' cultures such as Native Americans, Australian Aborigines and traditional African societies. Their spiritual systems (where they have been allowed to survive) are based on an intuitive sense of connection between all beings. Jain science retains that intuitive wisdom and acknowledges it as one of its most ancient roots. At the same time, it asks us to use our reasoning powers to understand our place in the cosmos and find our true selves, in other words find union with the Jiva or soul element within.

The insight that ecological awareness and spiritual practice are one and the same springs from intuition. It is not unique to the Jains and seems to be one of the earliest spiritual reference points for humanity. Interconnectedness is a pronounced feature of Shinto, for example, the indigenous religion of Japan which has roots as ancient as those of the Jains. It has also been powerfully expressed through many millennia by the Australian Aborigines and underlies every aspect of their spiritual practice. Jainism, however, is defined exclusively by the principle of interconnectedness. Every aspect of Jain life and thought arises from this principle and seeks to apply its lessons to the ethical dilemmas faced by individuals and the community as a whole.

Interfaith dialog and multiculturalism, like ecological or 'green' perspectives, are concepts that can also be seen to resemble certain areas of Jain thought. Nonetheless, the idea of Many-Sidedness is surely especially well qualified to help men and women today ('of all faiths and none') to navigate the complexities increasingly apparent within their own societies and in their relationship to others. It offers an alternative to the adversarial forms of politics that increasingly act as barriers to understanding between peoples or the workings of civil society. Many-Sidedness is also an outgrowth of Ahimsa (non-injury): an intellectual non-violence based on openness to ideas and respect for others.

This last quality explains the ease with which the Jain path takes us logically from intuitive to rational thought. At no stage are reason and intuition placed in opposition to each other or seen as rival perspectives. Their relationship is one of continuity and collaboration. Without reason, intuition becomes ethereal and loses its connection with reality. Equally dangerous is reason without intuition, which becomes abstract, arid and mechanical, losing its humanity and ultimately its truth. A partial analogy is provided by Yin and Yang in Daoist thought. The cosmos, and the human psyche, are made whole by the continuous interplay of the Yin (intuitive, 'soft') and Yang (rational, 'hard') elements. The interaction of Yin and Yang, so central to Chinese spiritual and aesthetic sensibility, expresses well the way relationship between spiritual and scientific processes should work, if they are to increase our understanding of ourselves and our compassion for all forms of life. However, in Jain thought, reason and intuition do not even engage in interplay, but journey together as part of the same process: the quest for knowledge.

We have already noted that Jainism contains many of the characteristics associated with 'indigenous' spiritual traditions. These are ways of thinking and living that are grounded in the experiences of peoples who interact constantly with natural forces. The Jain path starts from this point. According to its followers, it is the expression of the truly indigenous and primal Indic wisdom, before the development of caste and priestly hierarchies. Unlike the spiritual systems of the Native Americans and Australian Aborigines, for example, Jain thought has evolved freely, without distortion or suppression, adding new layers of thought and subtly adapting itself to social, economic and technological changes while maintaining its inner core. With the exception of Shinto in Japan, Jainism is probably the only indigenous tradition to survive in this uninterrupted form and to have adapted itself to the needs of a technologically advanced, culturally diverse society.

Over the past half century, the dialog between Christianity and secular science has proved to be increasingly creative and fruitful. Comparable processes are taking place globally. Thanks in large measure to the ground-breaking work of Fritjof Capra (*The Tao of Physics*), we have learned much about the relevance of Daoist and Buddhist thought to the physical sciences. Greater cultural openness – Many-Sidedness in action – has meant that across the world indigenous spiritual pathways are being reconstructed with dedication and care. Increasingly, it is apparent that their world views accord well with – and at times surpass – our most recent scientific advances and our present environmental concerns. In this context F. David Peat's *Blackfoot Physics* might well turn out to be a more significant study than Capra's better known book. Peat's research offers us an example of a physicist trained in the traditions of the European Enlightenment who is exploring the wisdom of a radically different North American culture and finding his own scientific understandings reinforced and then enlarged. As such, it is the basis for many of the assumptions behind our brief exploration of Jain science.

Until this point, the Jain tradition has been greatly underrepresented in the emerging science-spirituality dialog. There are two probable reasons for this, which are intertwined. First,

Jainism is not very well known, especially outside India, although significant Jain populations exist in East Africa, North America and Australia, with smaller communities in the United Kingdom and mainland Europe. Secondly, respect and admiration for the Jains exists because of their known emphasis on conserving life, but few of their doctrines are widely disseminated. Popular images of Jain ascetics, chiefly their extreme care in avoiding harm to small creatures, can confuse non-Jains.

The emergence of the 'animal rights' movement within western societies has created a high level of confusion about Jain doctrines. Expressed succinctly, animal rights theory posits equality between all forms of fauna, humans included. Animals exploited by humans are victims of 'species-ism', which is the equivalent of racism or sexism. All animals have inalienable rights and in this context differences in capacity and biological function between the species have no more relevance than differences of ability between humans. Therefore, human society needs to be reorganized to avoid or minimize the use of animals by humans for food, science, entertainment or even companionship.

Jain concepts differ from those of the animal rights movement in several important respects. The first and most obvious of these is the idea of a 'hierarchy of life' whereby life forms have different functions, abilities and capacities, which reflect different stages of spiritual as much as physical evolution. Jains acknowledge and celebrate evolutionary connections between species and the common possession of Jiva. The ecological connection between all living systems is, as we have seen, a key component of Jain science. Yet the hierarchy of life is expressed through a hierarchy of consciousness, so that Jains have no compunction about assigning a higher spiritual purpose to one life form than another.

Another critical difference between the two viewpoints is that animal rights ideology in the West is based on a materialistic conception of the person (animal or human) and the consequent reductionist view of the self. Thus the concept of personal autonomy is equated with *independence* rather than *interdependence*. Taken to its logical conclusion, this means that animal rights campaigners will often argue against any form of human 'interference' with nature, even when it involves using human intelligence and compassion rather than exploitation. The autonomy of other species may be placed above protecting them from becoming extinct or, in general terms, safeguarding biodiversity.

A typical example of this approach has been the release of mink, a non-indigenous species, into the British countryside by campaigners who raided fur farms. This act of political protest ignored the possible effects on other species, the threat to the natural balance and the ultimate welfare of the mink themselves! From a Jain standpoint, this is an example of careless action, arising from a one-sided viewpoint that fails to take account of ecological complexities.

Jains, by contrast, value every life in its own right *and* as part of something larger than itself. Self-realization involves setting the self in a wider context. Although Moksha as spiritual liberation promises the liberated soul release from all forms of action, the spiritually aware man or woman acts with compassion and seeks to protect life in all ways possible – by limiting harmful actions, but also through carefully considered positive acts. Rather than adopting a *laissez faire*, libertarian position towards the natural world, Jains seek a scientific understanding of natural processes and the place of humanity within them, seeing all of life as engaged in the same spiritual endeavor but operating at different levels.

This viewpoint sometimes leads to a convergence of aims with the animal rights movement. For example, Jains always look for positive and enriching alternatives to vivisection, factory farming and meat-eating, although their methods are always peaceful and based on the idea of following principles themselves rather than trying to impose them on others. The culture of Jainness involves continuous self-criticism and evaluation of one's own motives and equates fanatical or highly partisan positions with lack of understanding. The premises behind Jain philosophy and the animal rights movement are therefore quite different, even where some of their conclusions are superficially the same. Jainism does not extend the concept of rights across the whole spectrum of life, but confers responsibilities relative to levels of spiritual advancement. Human compassion towards other beings (including fellow men and women) is held to be evidence of humanity's distinction from other species. Human destructiveness (towards fellow humans and/or the rest of nature) is held to be a descent to a lower level of consciousness, reminiscent of less advanced life forms. This is so even when the destructive behavior involves ingenious and intricate forms of technology (weapons systems for example), the results of negative use of human intelligence. For humans, who are equipped with moral choice, the results are especially disastrous from the karmic perspective. This perspective has radical implications for human behavior and thought.

The western world is now, albeit anxiously, beginning to confront the social, psychic and above all ecological consequences of the breaking of the scientific-spiritual continuum. At the same time, India has emerged as a leading player on the global economic stage (see Mahapragya and Abdul Kalam, 2008). In this environment, the confluence of Indic and western thought is an intellectual exercise of especial power.

# Notes

## Introduction

1. This quotation is often attributed to George Bernard Shaw (see Appendix 2), but it is unclear where, or even whether, he used these exact words. What is known is that he admired Jain teachings on respect for life, including dietary habits. In a conversation with Mr Devadas Gandhi, son of the Mahatma, that took place circa 1950, 'George Bernard Shaw … expressed his view that the Jain teachings were appealing to him much and that he wished to be born after death in a Jain family. Due to the influence of Jainism he was always taking pure food free from meat … and liquors.' (Vijay, M.K., translated by Badami, P.S. (1954), *Jainism in a Nutshell*. Jain Gyana Mandir, Bombay. 71)
2. The quotation is from the scripture *Bhaktapragya*, verse 93.
3. This approach differs distinctly from the ideology of 'animal rights' in the West, with which it is at times conflated. Animal rights theory postulates political equality for all species (or a status approximating to this) and tends to emphasize the autonomy of the individual being, rather than mutual responsibility or hierarchy (see Appendix 2).

## Chapter 2

1. For a detailed discussion of the social and philosophical implications of *Anekantavada* see Rankin, Aidan (2010) *Many-Sided Wisdom: A New Politics of the Spirit*. O-Books, Winchester/Washington DC. Also, Rankin, Aidan (2006) *The Jain Path: Ancient Wisdom for the West*. O-Books, Winchester/Washington DC. 159-193
2. Mahalanobis, P.C. (1954) 'The Foundations of Statistics', *Dialectica* 8. 95-111

## Chapter 3

1. Jinasena, 'There is No Creator'. *Mahapurana,* as quoted in Sproul, Barbara C. (1991) *Primal Myths: Creation Myths around the World*. HarperCollins, San Francisco. 92
2. Jaini, P.S. (1979): *The Jaina Path of Purification*. Berkeley: University of California Press (Reprinted by Motilal Banarsidass, Delhi). 114. 'Jainas speak of the "innumerable qualities" of the soul. Nevertheless, it can legitimately be said that the presence of those qualities which have been briefly discussed above – perception, knowledge, bliss, and energy – are sufficient to define the soul as a totally distinct and unique entity, an existent separate from all others.'
3. *ibid.* 113
4. *ibid.* 112

## Chapter 4

1. Jaini (1979). 109

## Chapter 5

1. Jaini (1979). 125. 'Nama-karmas pertaining to sarira [*or sharira*] are also said to generate two subtle bodies underlying the manifest physical one. These are the *taijasa-sarira,* heat body, which maintains the vital temperature of the organism, and the *karmana-sarira,* the karmic body, constituting the sum total of karmic material present in the soul at a given time. The conception that such bodies exist is important to the Jain theory of rebirth, since they constitute the 'vehicle' whereby a soul moves (albeit under its own power) from one incarnation to the next.'
2. *ibid.* 126-7. 'At the moment of death, the *aghatiya* karmas have pre-programmed, as it were, the particular conditions of the coming embodiment. This information is carried in the karmana-sarira, which together with the taijasa-sarira, houses the soul as it leaves its physical body. A soul is said to be

inherently possessed of great motive force; set free of the state of gross embodiment, it flies at incredible speed and in a straight line to the destination which its accompanying karma deemed appropriate. This movement is called vigraha-gati, and it is said to require, as noted above, only a single moment in time, regardless of the distance to be traversed.'

3. *ibid.* 98. 'The distinguishing quality of space is its ability to provide a locus for such existents; this is true whether it actually does so (as in the case of loka-akasha) or not (as in the case of aloka-akasha). Hence, there is only one "space"; its extent is infinite. Akasha is further described as divisible into infinitesimally small "space-points" (*pradesha*); these units have some dimension and yet cannot be sub-divided.'

## Chapter 6

1. Jaini (1979). 112. 'The energy quality, 'perverted' by this impurity, produces vibrations (*yoga*), which bring about the influx (*asrava*) of different kinds of material karma. The vibrations referred to here actually denote the volitional activities of the individual. Such activities can be manifested through either body, speech, or mind; ...'
2. *ibid.* 113. 'The precise amount (*pradesha*) of karma that engulfs the soul after a given activity is said to depend upon the *degree of volition* with which that activity was carried out. The type of activity, moreover, determines the specific nature (*prakruti*) assumed by the theretofore undifferentiated karmic matter. ... As for the duration *(sthiti)* and result *(anubhava)* of given karmas – how long they will cling to the soul and what precise momentary effect they will eventually have upon it – these are fixed by the degree to which such passions (*kashaya*) as anger and lust colored the original activity. Once a karma has given its result, it falls away *(nirjara)* from the soul "like ripe fruit", returning to the undifferentiated state and thus to the infinite pool of "free" karmic matter; ...'

## Chapter 7

1. Jaini (1979). 169. '...the dead flesh itself is a breeding ground for innumerable nigodas* and hence must not be consumed.' (* 'Nigodas': Professor Jaini here uses an anglicized plural form, adding an 's'. In *Living Jainism,* we have broadly adhered to the traditional Sanskrit or Prakrit forms, which usually render the plural in the same way as the singular. Within Indic studies, both the traditional and the anglicized renderings are considered equally 'correct'. See Jain Glossary.)
2. *ibid*. 168. 'Such creatures (*nigoda*) are said to be especially prevalent in substances where fermentation or sweetness is present; hence the consumption of liquor or honey brings untold millions of these organisms to an untimely and violent end. The tissues of certain plants, especially those of a sweet, fleshy, or seed-filled nature, are also thought to serve as hosts for nigoda; plants of this type are termed sadharana, "those which share their bodies". The avoidance of figs as part of the mulaguna practice seems to represent a symbolic renunciation of all nigoda-ridden vegetable substances; ...'
3. *ibid*. 171. 'A murderer, for example, clearly sets out to end the life of his victim, hence commits samkalpaja-himsa. Surgeons, on the other hand, may cause pain or even death during a delicate operation, but are guilty only of the much less serious arambhaja-himsa.'
4. *ibid.* 32. '...at every moment there is a living Jina *somewhere*. In other words the path of salvation is open at any time.'

## Chapter 8

1. Jaini (1979). 140-1 '...thanks to the fluctuations in the ongoing interaction of virya (*energy*) and karma, certain experiences (especially an encounter with a Jina or his image, hearing the Jaina teachings, or remembering past lives) *may* bring the bhavyatva out of its dormant state and thus initiate the

process that leads eventually to Moksha.'

2. *ibid.* 147. 'Previously he has identified his being in external signs of life – the body, states, possessions; thus he has been in the state known as *bahiratman,* seeing the self in externals dominated by the consciousness which is aware only of the results of karma *(karma-phalacetana)...* This orientation depends on the false notion that one can be the agent *(karta)* of change in other beings; ...'
3. *ibid.* 150. 'This awareness of the basic worth of all beings, and of one's kinship with them, generates a feeling of great compassion *(anukampa)* for others. Whereas the compassion felt by an ordinary man is tinged with pity or with attachment to its object, anukampa is free of such negative aspects; it develops purely from wisdom, from seeing the substance (*dravya*) that underlies visible modes, and it fills the individual with an unselfish desire to help other souls towards Moksha.'
4. *ibid.* 159. 'In the last few moments of embodiment, even yoga is brought to cessation; this state of utter immobility is called omniscience without activities *(ayoga-kevalin),* the fourteenth gunasthana. At the instant of death (nirvana) itself, the soul is freed forever from the last vestige of samsaric influence; ...'

## Chapter 9

1. Jaini (1979). 252-3. 'Dharmadhyana entails the intense contemplation, for a short period (up to forty-eight minutes), of one of several objects: (1) the teachings of the Jina on the nine tattvas and how these teachings can best be communicated to others (*ajnavicaya*); (2) the great misery suffered by other beings (whose minds are impelled by passions and blinded by ignorance) and the means by which these beings can be saved (*apayavicaya*); (3) the mysterious mechanisms of karmic influx, binding, duration, and outcome and the fact that the soul is fundamentally independent of these processes and thus able

to disengage itself therefrom (*vipakavicaya*); (4) the structure of the universe and the interplay of causes that brings souls to their particular destinies (*samsthanavicaya*).'

## Epilogue

1. *Nature* (1940). 605. The same aphorism was subsequently used in a letter to the Jewish Philosopher Eric Gutkind, 3 January, 1954. This was written after receipt of Gutkind's book, *Choose Life: The Biblical Call to Prayer* (1952, first published 1922 or 1923). In it, Einstein goes on to reject many of the specific tenets of the Jewish religion, while approving of its underlying values and arguing that religious faith is an important part of the human condition. Einstein was offered the opportunity of becoming the second President of the State of Israel, but he declined in order to devote all his time to scientific research and writing.
2. Quoted in Peat, David *Blackfoot Physics* (2006), Weiser, York Beach ME, 140-1. See also Bohm, David, *Wholeness and the Implicate Order* (2002), Routledge, London.
3. The story of the elephant exists in several slightly different versions amongst Hindus and Sufis as well as Jains. This is, in itself, an example of Anekantavada, or Many-Sidedness.
4. For the definitive work on Gaia theory, see Lovelock, James *Gaia: A New Look at Life on Earth*, Oxford Paperbacks, Oxford, 2000.
5. See Devall, Bill, Naess, Arne and Drengson, Alan (eds) *The Ecology of Wisdom: Writings By Arne Naess*, Counterpoint, Berkeley, 2010.

## Appendix 2

1. Mahapragya, Acharya and Abdul Kalam, Avul Pakir Jainulabdeen (2008) *The Family and the Nation*, HarperCollins, New Delhi.

# Further Reading

## Principal Jain Texts

Jacobi, Hermann (1968 reprint) *Kalpa-sutra*, Dover Publications, New York (included in *Jaina Sutras, Volume* 2, see 'General' section below)

Quarnstrom, Olle (2000) *The Yogasutra of Hemacandra: A Twelfth Century Handbook of Svetambara Jainism*, Harvard University Press, Cambridge, MA

Singhvi, S.L. (tr.) and Jain, Vardhman Sthanakvasi (tr.) (1984) *Karmagrantha of Devedrasuri, Parts 1-6*, Dharmic Siksha Samiti, Meerut (in Hindi)

Tatia, Nathmal, *That Which Is* (*Tattvartha-sutra of Umasvati*) Harper Collins, San Francisco 1994 (part of the 'Sacred Literature' series)

## General

Basham, A.L. (1958) 'Jainism and Buddhism', in de Bary, W.T., *Sources in Indian Tradition, Vol 1*. Columbia University Press, New York. 38-92

Bohm, David (2002) *Wholeness and the Implicate Order*, Routledge, London

Capra, Fritjof (1975) *The Tao of Physics*, Wildwood House, Hounslow

Carrithers, Michael and Humphrey, Caroline (1991) *The Assembly of Listeners: Jains in Society*, Cambridge University Press, Cambridge (UK)

Chitrabhanu, Gurudev Shree (1980) *Twelve Facets of Reality: The Jain Path to Freedom*, Dodd, Mead & Co., New York

Davies, Paul (1983) *God and the New Physics*, J.M. Dent and Co., London

Devall, Bill (ed.), Naess, Arne (ed.) and Drengson, Alan (ed.) (2010) *The Ecology of Wisdom: Writings By Arne Naess*,

Counterpoint, Berkeley
Dundas, Paul (1992) *The Jains*, (2nd edition) Routledge, London
Einstein, Albert (1940) 'Science and Religion,' *Nature*, Vol. 146. 605-7
Glasenapp, H. von (1942) *The Doctrine of Karman* (Karma) *in Jain Philosophy*, P.V. Research Institute, Varanasi
Gribbin, John (1984) *In Search of Schrodinger's Cat*, Wildwood House, Hounslow (reprinted by Corgi Books)
Haldane, J.B.S. (1957) 'The Syadvada System of Predication,' *Sankhya*, no. A18. 195-200
Hawking, Stephen (1988) *A Brief History of Time*, Bantam Press, London
Hay, S.N. (1970) 'Jain Influences on Gandhi's Early Thought' in Ray, S. (ed.) *Gandhi, India and the World*, Temple University Press, Philadelphia. 29-38
Jacobi, Hermann (1968) *Jaina Sutras, Vols 1 and 2*, Dover Publications, New York (Reprint of 1884 and 1895 editions)
Jain, C.R. (1929) *The Practical Dharma*, The Indian Press, Allahabad
- (1974) *Fundamentals of Jainism*, Veer Nirvan Bharti, Meerut
Jain, S.K. (1980) 'Communication Regarding the Process of Rebirth' in O'Flaherty, Wendy Doniger (ed.) *Karma and Rebirth in Classical Indian Traditions*, University of California Press, Berkeley. 237-8
Jaini, J.L. (1971) *Outlines of Jainism* L. Jaini Trust, Indore
Jaini, Padmanabh S. (2001) *The Jaina Path of Purification*, Motilal Banarsidass, Delhi
Kachhara, N.L. (2005) *Jaina Doctrine of Karma*, Dharam Darsham Sewa Samthan, Udaipur
Lovelock, James (2000) *Gaia: A New Look at Life on Earth*, Oxford Paperbacks, Oxford
Mahalanobis, P.C. (1954) 'The Foundations of Statistics', *Dialectica* 8. 95-111
Mardia, Kanti V. (1975) 'Jain Logic and Statistical Concepts', *Jain*

*Antiquary and Jaina Siddhanta Bhaskar*, Oriental Research Institute, Arrah, Bihar

- (1976) 'Do-It-Yourself Statistical Analysis,' *Leeds University Review*, no. 19. 79-98

- (1981) 'Why Paryushana is Doing Your Own MOT,' *The Jain*, no. 3, Issue 9. 4-5*

- (1982) 'Mahavira As a Man,' *The Jain*, no. 4, Issue 11. 16

- (1988a) 'Probability, Statistics and Theology' (Discussion with Bartholomew, D.J.), *Journal of the Royal Statistical Society*, no. 151. 166-7

- (1988b) 'Jain Culture,' *The Jain*, Pratistha Mahotsava Souvenir Issue. 116-19

- (1990) *The Scientific Foundations of Jainism* (1st edition), Motilal Banarsidass, Delhi

- (1991) 'Modern Science and the Principle of Karmons in Jainism,' *Jain Journal*, Vol. 26.116-19

- (1992) *Jain Thoughts and Prayers*, Yorkshire Jain Foundation, Leeds

- (2007) *The Scientific Foundations of Jainism* (2nd edition), Motilal Banarsidass, Delhi

(Hindi Translation: (2004) *Jain Dharma ki Vigyanik Adharshila*, Parsvanath Vidyapitha, Varanasi)

(Gujarati Translation: (2011) *Jain Dharmani Vigyanik Adharshila*, L.D. Institute of Indology, Ahmedabad)

- (2008) 'Modern Science and the Four Noble Truths of Jains,' *Young Jaina International Newsletter*, Watford (UK), Vol. 22 (1), February-May 2008

- (2009) 'Knowledge-based Jainism & Eco-Warriors', Ecology – The Jain Way: 15th Biennial JAINA Convention, Los Angeles, 2009. 44-5

Marett, Paul (1985) *Jainism Explained*, Jain Samaj Europe Publications, Leicester

Mehta, Mohan Lal (2002) *Jaina Psychology: An Introduction*, Parsvanath Vidyapeeth, Varanasi

Pal, P. (1994) *The Peaceful Liberators: Jain Art from India,* Thames and Hudson with Los Angeles County Museum of Art, Los Angeles

Peat, F. David (2006) *Blackfoot Physics: A Journey into the Native American Universe,* Weiser, York Beach, ME/Boston

Popper, Karl R. (1968) *The Logic of Scientific Discovery* (Second Edition), Hutchinson, London

Ramchandran, G.N. (1982) 'Syad Naya System (SNS) – A New Formulation of Sentential Logic and its Isomorphism with Boolean Algebra of Genus 2', *Current Science,* Vol 51. No 13. 625-36

Rankin, Aidan (2006) *The Jain Path: Ancient Wisdom for the West,* O-Books, Winchester/Washington DC

- and Shah, Atul K. (2008) *Social Cohesion: A Jain Perspective,* Diverse Ethics Ltd, Colchester

- (2010) *Many-Sided Wisdom: A New Politics of the Spirit,* O-Books, Winchester/Washington DC

Shah, Atul K. (1991) *Experiments with Jainism,* Young Jains Publications, London

- and Rankin, Aidan (2008) *Social Cohesion: A Jain Perspective,* Diverse Ethics Ltd, Colchester

Sheldrake, Rupert (1983) *A New Science of Life,* Blond and Briggs Ltd, London (reprinted by Paladin, 1983)

Singhvi, L.M. (1990) *The Jain Declaration on Nature,* The Jain Sacred Literature Trust, London

Stevenson, Sinclair (1970) *The Heart of Jainism,* Motilal Banarsidass, Delhi (reprint of Oxford University Press edition, 1915)

Tobias, Michael (2000) *Life Force: The World of Jainism,* Jain Publishing Company, Fremont, CA

Zaveri, J.S. (1975) *Theory of Atom in the Jaina Philosophy,* Jaina Vishva Bharati, Ladnun

**The Jain* magazine was published by Jain Samaj Europe in Leicester (UK)

# Jain Glossary

This section aims to provide the reader with a wide-ranging list of Prakrit and Sanskrit terms used by Jain practitioners to describe the intellectual and spiritual world view of Jainism. Many of these terms are found in our text. We have attempted to provide as comprehensive a reference as possible, although it is impossible to encompass the diversity of Jain thought and practice, which evolves continuously in the many areas of the world where Jain communities now exist.

The English translations are in many cases approximate equivalents. In the case of key concepts or inner teachings, we have rendered the translations in the upper case (e.g. Tri-Ratna, Three Jewels). Otherwise the lower case is used (e.g. Deva, heavenly being). In the interests of clarity and neatness, we have removed diacritical marks but we have remained as faithful as possible to the Sanskrit and Prakrit scripts in which most of these words were originally rendered. However, in certain cases, more familiar renderings exist (e.g. Moksha rather than Moksa) and we have adhered to these to avoid confusion.

We have generally adhered to the traditional Sanskrit or Prakrit grammatical formations, which render the singular and plural in the same way, e.g. *Guna,* quality or qualities. In some instances, we have used an anglicized plural form when that form is widely accepted and used by Anglophone practitioners or students of Jain Dharma. For example, there are occasions in the book where we refer to the nine *tattva* ('reals', aspects of reality) as '*tattvas*', where it makes more immediate sense or reads better.

Acharanga, a Jain scripture
Acharya, spiritual master
Adharma, dynamic medium, a Dravya (q.v.)
Agamas, Jain scriptures

Ahimsa, Non-Violence / non-injury / avoidance of harm
Ajiva, non-soul/insentient object
Akasha, Space
Aloka-Akasha, unoccupied space
Anekantavada, Jain holistic principle, principle of 'many-sidedness'
Anga, Jain scriptures: main texts (literally 'limbs')
Angabahya, Jain scriptures: subsidiary texts
Anitya, impermanence
Anubhava, potential energy in karmon-decay
Anupreksha, reflections – there are twelve forms
Anuvrata, Five Lesser Vows
Anuyoga, secondary scripture
Anyatva, separateness
Aparigraha, non-possession, non-possessiveness
Arihanta, perfect being
Avadhijnana, clairvoyance
Avasarpini, regressive half-cycle
Avirati, non-restraint
Ayoga-kevalin, static omniscience state
Ayu (karmic component, secondary, longevity determining)

Bandha, karmic bondage / volition
Bhava, volition
Bhavyatva, freedom-longing catalyst

Darshana, perception
Dasavaikalika, a scripture
Deva, heavenly being
Dharma, righteousness, universal law, universal order
Dhyana, meditation
Digambara, one of the two main schools of Jainism: literally means 'sky-clad' because the male ascetics are naked
Dravanuyoga, exposition on the existents: includes Tattvartha-

sutra (q.v.)
Dravya, the Six Existents
Dvesha, aversion
Ekant, one-sidedness, doctrinaire viewpoint

Gati, Four Existences: Deva (heavenly being) (q.v.); Manusya or Manushya (human) (q.v.); Naraki (hellish being) (q.v.) and Tiryancha or Tiryanka (animal or plant life) (q.v.)
Ghatiya, primary karmic components
Gotra, environment determining karmic component
Guna, element of the soul
Gunasthana, fourteen purification stages
Gupti, restraint

Himsa, violence, harm, destructive power

Irya Samiti, Careful Action

Jai Jinendra, 'Honor to (the) Jina', 'Hail to the Conqueror', popular Jain greeting
Jina, spiritual victor, omniscient spiritual teacher (see also Tirthankara)
Jiva, soul, living being, that which is alive
Jnana, knowledge
Jnana-avaraniya, knowledge-obscuring karmic component

Kala, time
Kala (2), temporal cycles
Karma, subtle matter composed of karmic particles, attracted to the Jiva (q.v.) by yoga (activity) (q.v.) and preventing full self-knowledge, omniscience and transcendence of cycle of birth, death and rebirth
Karma (2), any of the eight karmic components
Karmic Sharira, types of bodies (see Sharira)

Karmon (pl. Karmons), karmic particles (modern scientific adaptation of traditional Jain concept)
Kashaya, Passions (Four Principal Passions: Anger = krodha; pride = mana; deceit = maya; greed = lobha
Kaya, body
Kevaljnana, infinite omniscience

Leshya, karmic 'stain' on Jiva/soul; 'coloration' of Jiva/soul reflecting karmic influence
Loka Akasha, unoccupied space
Lokakasha, inhabited universe occupied space

Mana, mind
Mana (2), pride
Manusya, human
Matijnana, empirical knowledge
Maya, deceit
Mishra, mixture of deluded and enlightened world views
Mithyadarshana, perverted or distorted world view
Mohaniya, bliss-defiling karmic component
Moksha, spiritual liberation, (moment of) enlightenment, acquisition of omniscience, liberation from cycle of death and rebirth (Samsara)

Nama, body producing karmic component
Naraki, hellish being
Naya, standpoint
Nayavada, unique standpoint principle
Nigoda, micro-organism(s)
Nirjara, karmic fission/decay
Nirvana, release from (physical) bondage, final physical death of enlightened man or woman, followed immediately by Moksha
No-Kashaya, secondary or subsidiary passions

Panca-paramesthin, five spiritual heights
Papa, heavy karmic matter
Paramanu, ultimate particle (UP)
Parasparopagraho Jivanam, all life is interconnected / bound together
Parigraha, possession, possessiveness
Prabhavana, illumination
Pradesha, space point
Pradesha (2), number of karmons
Prakruti, karmic components of karmic force.
Pramada, carelessness
Pramana, comprehensive right knowledge
Pudgala, matter, phyiscal energy
Punya, light karmic matter
Purva, Jain scriptures (old texts)

Raga, attachment
Rutradhyana, wrathful meditation

Sadhu, saint
Samiti, watchfulness, care
Samsara, cycle of physical incarnation: birth, death and rebirth
Samvara, karmic force shield
Samyak-charitra, Right Conduct
Samyak-darshana, Right Faith
Samyak-jnana, Right Knowledge
Sata-vedaniya, karmic component, pleasure producing
Satya, truth, truthfulness
Sharira, types of (karmic) bodies
Sharira (2), karmic body
Sharira (3), karmic capsule
Shramana, ascetic(s); monk(s) and nun(s); laborers on the spiritual path
Shravaka, Jain lay man

Shravika, Jain lay woman
Siddha, liberated soul
Sukha, bliss
Sukladhyana or Shukladhyana, pure trance
Susama, happiness
Svetambara, one of the two main schools of Jainism: literally means 'white-clad' because the male and female ascetics wear white
Syadvada, conditional predication principle
Syat, maybe, perhaps, expression of possibility

Taijas sharira, karmic capsule (see Sharira)
Tattva, Tattva(s), Nine Reals
Tattvartha-sutra, 'That Which Is': summary of whole Jain doctrinal system by Umasvati, Second Century CE
Tirtha, the order of monks, nuns, lay men and lay women
Tirthankara, omniscient spiritual teacher
Tiryancha (sometimes Tiryanka), animal or plant life
Tri-Ratna, Three Jewels of Jainism: Samyak-darshana (Right Faith) (q.v.); Samyak-jnana (Right Knowledge) (q.v.) and Samyak-charitra (Right Action/Conduct) (q.v.). Also known as Ratnatraya

Udaya, emission of karmic particles
Upadhayaya, spiritual teacher
Utsarpini, progressive half-cycle, see also Kala

Vachan, speech
Vargana, particle-groupings, variforms
Vatsalya, disinterested love
Vayu-kayika, air bodies
Veda(s), sacred texts of the Hindu tradition, grouped in four canonical collections: Rigveda, Yajurvedam Samaveda and Atharvaveda. Veda is Sanskrit word for knowledge.

Vedaniya, feeling-producing karmic component
Virodhi-himsa, defensive violence
Virya, energy
Virya-antaraya, energy obstructing karmic component

Yoga, activities of body, mind and speech

# Index

MANTRA
BOOKS

Printed and bound by CPI Group (UK) Ltd, Croydon, CR0 4YY